Environmental Risk, Environmental Values, and Political Choices

Environmental Risk, Environmental Values, and Political Choices

Beyond Efficiency Trade-offs in Public Policy Analysis

EDITED BY
John Martin Gillroy

Routledge
Taylor & Francis Group
LONDON AND NEW YORK

First published 1993 by Westview Press, Inc.

Published 2018 by Routledge
52 Vanderbilt Avenue, New York, NY 10017
2 Park Square, Milton Park, Abingdon, Oxon OX14 4RN

Routledge is an imprint of the Taylor & Francis Group, an informa business

Library of Congress Cataloging-in-Publication Data
Environmental risk, environmental values, and political choices : beyond efficiency trade-offs in public policy analysis / edited by John Martin Gillroy.
p. cm.
Includes bibliographical references and index.
ISBN 0-8133-8147-9
1. Environmental policy—Congresses. 2. Human ecology—Moral and ethical aspects—Congresses. 3. Technology—Risk assessment—Social aspects—Congresses. I. Gillroy, John Martin, 1954– .
HC79.E5E586 1993
363.7—dc20 **92-45788**
CIP

ISBN 13: 978-0-367-01612-8 (hbk)

ISBN 13: 978-0-367-16599-4 (pbk)

Contents

Acknowledgments

This collection of essays has grown from discussions among the authors on the subjects of philosophy, public policy, and environmental risk. These discussions were aided by the meeting of most of the contributors at a symposium held on the campus of Trinity College on March 16, 1990. Entitled "Environmental Policy in the '90s and Beyond: Science, Human Values, and Political Choices," the symposium was partially underwritten by Trinity's New Liberal Arts Grant from the Sloan Foundation. At this meeting it was jointly decided to express our concerns on this subject by writing essays that could be gathered together in book form.

The editor wishes to acknowledge the Sloan Foundation, J. Ronald Spencer, Andy Gold, Dan Lloyd, Glen Gross, Paul Diaz, and Steven Jud for their tireless help in putting together the symposium that encouraged the writing of these essays. I would also like to thank my wife, Margaret Murray, for her patience and editorial assistance; Amy Eisenberg of Westview Press for her faith in this project; and, most especially, Marilyn Brazil, for her help in the preparation of this book. Without Marilyn the task could not have been accomplished.

John Martin Gillroy

About the Contributors

Christopher J. Bosso is associate professor and chair of the Department of Political Science at Northeastern University, Boston, Massachusetts. He is the author of *Pesticides and Politics: The Life Cycle of a Public Issue* and has current research interests in public policy processes (particularly agenda setting) and environmental interest group formation.

John Martin Gillroy is assistant professor and Dana Faculty Fellow in the Department of Political Science and director of the Public Policy Studies Program at Trinity College in Hartford, Connecticut. His primary interest is in the relationship between (normative and positive) political thought and public policy choice, with special emphasis on the use of Immanuel Kant's ethical and political theory in the formulation and prosecution of policy argument.

William C. Gunderson is professor of political science at Carthage College in Kenosha, Wisconsin. Since pursuing his doctoral research in the Korhogo Region of the Ivory Coast in the early 1970s, he has been interested in center-periphery relations. During the 1980s, he spent a sabbatical leave and four years on the Racine County Board of Supervisors examining the implementation of national and state policy at the county level. In 1990, he moved to Montpellier, France, where he examined changes in local politics given the early 1980s' decentralization reforms.

David E. Henderson is professor of chemistry at Trinity College in Hartford, Connecticut, and director of the college's Interdisciplinary Science Program. He has written on air pollution for the *World Book Encyclopedia* and has recently published a monograph entitled *Air Pollution and Risk Analysis.* He is currently developing a program to allow chemistry students at a number of colleges in the northeast to monitor acid rain as a part of their general chemistry laboratories.

Steven Kelman is professor of public policy studies at the John F. Kennedy School of Government, Harvard University, Cambridge, Massachusetts. He is author of many books and essays, including

What Price Incentives? Economists and the Environment, a seminal treatise on the use of the principle of efficiency in making environmental policy decisions.

Douglas MacLean is presently chairman of the Philosophy Department at the University of Maryland in Baltimore. Formerly he was director of the Institute for Philosophy and Public Policy at the University of Maryland, College Park. He is editor of *Values at Risk* and author or editor of many other books and essays on policy matters.

Vicente Medina is assistant professor of philosophy at Seton Hall University, South Orange, New Jersey. His interests include social contract theory and the relationship between environmental policy and moral values.

Robert C. Paehlke teaches in the Political Science Department and the Environmental and Resource Studies Program at Trent University, Peterborough, Ontario, Canada. He is the founding editor of the journal *Alternatives: Perspectives on Society, Technology and Environment*. His recent publications include *Environmentalism and the Future of Progressive Politics* and *Managing Leviathan: Environmental Politics and the Administrative State.*

Talbot Page is professor of economics at Brown University, Providence, Rhode Island. His studies have included environmental regulation, policy analysis, decision theory, normative economics, moral philosophy (amateur standing), incentive compatibility, game theory and its "rationality" foundations, and epidemiology of drinking water, all of which relate to the understanding and control of environmental problems. He is author of *Conservation and Economic Efficiency*.

Barry G. Rabe is associate professor of health politics in the School of Public Health and adjunct assistant professor in the Department of Political Science at the University of Michigan, Ann Arbor. He is the author of *Fragmentation and Integration in State Environmental Management* and coauthor of *When Federalism Works*. He is currently researching North American hazardous waste policy and environmental regulatory integration in the Great Lakes Basin.

Introduction

John Martin Gillroy

The purpose of this collection is twofold. First, we argue that environmental risk, as a policy problem, requires moving beyond the market principle of efficiency as the basis of decision making and toward the articulation and use of environmental values to produce good public choices. Second, we examine the conflicts between environmental values and other competing value systems, especially concerning the instrumentalist values of the market paradigm.

Environmental values as discussed herein are generally perceived as an alternative set of moral imperatives with the integrity or intrinsic worth of human beings and natural systems as a basic critical concern. Specifically, Robert Paehlke in *Environmentalism and the Future of Progressive Politics* (New Haven, Conn.: Yale University Press 1989, 144–45) has compiled a list:

1. An appreciation of all life forms and a view that the complexities of the ecological web of life are politically salient.
2. A sense of humanity regarding the human species in relation to other species and the global ecosystem.
3. A concern with the quality of human life and health, including an emphasis on the importance of preventative medicine, diet, and exercise to the maintenance and enhancement of human health.
4. A global rather than a nationalist or isolationist view.
5. Some preference for political and/or population decentralization.
6. An extended time horizon—a concern about the long-term future of the world and its life.
7. A sense of urgency regarding the survival of life on earth, both long-term and short-term.
8. A belief that human societies ought to be reestablished on a more sustainable technical and physical basis and an appreciation that many aspects of our present way of life are fundamentally transitory.

9. A revulsion toward waste in the face of human need (in more extreme forms, this may appear as asceticism).
10. A love of simplicity, although this does not include rejection of technology or "modernity."
11. An aesthetic appreciation for season, setting, climate, and natural materials.
12. A measurement of esteem, including self-esteem and social merit, in terms of such nonmaterial values as skill, artistry, effort, or integrity.
13. An attraction to autonomy and self-management in human endeavors and, generally, an inclination to more democratic and participatory political processes and administrative structures.

Although there is general consensus on the above values, the list is not comprehensive and could not be while the primary analytic and methodological approach of policy analysis within the United States remains fixed within the context of efficiency, instrumental values, and the market paradigm.

We make no claim that an authoritative environmental system of values exists or is operational within the policy process. The purpose of this collection is to suggest that it is necessary to move beyond the market paradigm and to organize and argue environmental policy from the foundation of environmental values. We hope these essays help readers to think about this possibility and begin the integration of these values and their resultant arguments into the mainstream of policy discourse.

All of the contributors to this volume understand that a risk-free society is not possible. We are suggesting that one ought to be able to judge risk ex ante and make some decisions, individually and collectively, about what is acceptable and for what reasons. To accomplish this we must decide what is critical about a risk situation and then define our policy within these terms. We contend that this cannot be accomplished as long as efficiency remains the primary policy imperative.

It is not instrumental efficiency that is most critical to environmental policy decisions but the intrinsic value of mankind and environmental systems. By "intrinsic" value we mean that value which is placed on something that is not instrumental to the achievement of something else. The human and moral integrity of the person, the functional integrity of the environment, and those material conditions that support human beings and the environment should be segregated and given a special place in the analysis and methodological investigation of environmental risk. However, this volume does not go into depth as to the meaning or various definitions of intrinsic value that could be applied to humans or the various parts of the environment. Rather, we hope the discussion herein leads to a dialogue about how these terms and ideas should be operationally defined for use in making public choices. All we assume is that there is a difference between the value of something as an end in itself and the value it has as an instrument to obtain something else.

The roots of the distinction between intrinsic and instrumental value and the need for a reorientation to a system of environmental values in risk policy are the subject of Part 1 of the book. The three chapters in Part 1 concern the existence of a replacement for market values with environmental values in the case of environmental risk (Gillroy), the historical roots of the distinction between market and non-market domains of value (Kelman), and the translation of environmental values into practical policy (Paehlke).

The value trade-offs and problems of operationalizing a system of environmental values are the concerns of Parts 2 and 3 of this volume. In Part 2, we examine the status of environmental values as a philosophical system (Medina) and how environmental values might or might not conflict with the institutions of the U.S. policy process (Bosso) or the values of scientific analysis (Henderson). In Part 3, we take as a case the NIMBY ("not in my back yard") problem, which we characterize as an instance of environmental risk regulation and prevention, examining its roots in the combination of power politics and the push for economic growth (Gunderson). Here the market paradigm seems challenged by a situation where economic incentives play a minor role. In an effort to examine solutions to the NIMBY situation, we present two alternative ways of conceptualizing the problem: from within the market system of values (Page) and from the standpoint of environmental values (Rabe and Gillroy).

In the Epilogue, MacLean reminds us of the conflict and compromise necessary in environmental policy and discusses the inherent tensions between the two imperatives that motivate public choice in this issue area: preservationism and conservationism. This "last word" is a reminder that with the introduction of environmental values into policy decision making, complexity is added to the choice, but a complexity that reflects the intricacy of humanity and nature, the result of which may be better collective choices and consequences.

Overall this book is a preliminary investigation of the need for and conflicts generated by the use of environmental values rather than market values to analyze public choices concerning environmental risk. We hope it will generate discussion and controversy and produce a higher level of argument about alternatives to the market paradigm, which, as we know, is not neutral but transports its own set of values into the analysis of public policy.

PART ONE

Value Domains, Integrity, and Policy Argument

1

Integrity, Intrinsic Value, and the Analysis of Environmental Risk

John Martin Gillroy

Public policy choice regarding humanity's relationship with the natural environment is concerned with human life and health, the value of natural resources, and the quality of the ecology as a dynamic life-support system, as well as with aesthetic considerations. The predominant approach to the evaluation of policy in this area places emphasis on the methodology of cost-benefit analysis and the principle of market efficiency. The purpose of this chapter is to examine the theoretical framework of environmental risk policy to see if the justification and evaluation structure of economic thought is morally adequate to the task of finding the "socially better" policy alternative in this area of concern.[1]

We will argue that, first, the logical and strategic structures promoted by cost-benefit methods do not provide for the collective good of environmental quality. Second, we will contend that the valuing scheme promoted by the application of the principle of efficiency to public choice does not and cannot address the intrinsic values at stake in the regulation of environmental risk. Third, we will suggest that only the hegemony of environmental values as a pre-efficiency test for environmental risk policy can provide an adequate foundation for policy decisions in this area.

Certain questions motivate this chapter. In the attempt to find an efficient result based on individual wealth preferences, are cost-benefit methods sufficient to handle the ethical ramifications of environmental policy? Is efficiency a morally adequate principle to inform policy decision making in this area? Can efficiency and welfare considerations properly value the integrity of humans or their natural environment? The dilemma will concern whether one can adequately compensate for involuntary transfer of external effects (Page 1973; Pigou 1932) by treating them as renegade market considerations within a value system based on instrumental economic efficiency, or whether they ought to be considered properly as collective action problems, external to markets and handled separately by a distinct set of environmental values.

Before we can come to a conclusion about the need for environmental values that can accommodate the special moral and strategic character of environmental risk problems, we must first describe the economic paradigm, examine the strategic ramifications of traditional cases of pollution, and consider the distinctive institutional, strategic and moral nature of environmental risk. Throughout, we will examine how the cost-benefit approach measures up to the demands of the environmental dilemma and assess its capacity to prescribe both the moral solutions and the political institutions necessary to respond to environmental risk situations. Finally, we will conclude that the ethical poverty of the economic approach requires a new direction in the justification and evaluation of environmental risk policy and a reorientation of the values at the foundation of this policy.

The Economic Paradigm

Economic reasoning is concerned primarily with efficient voluntary exchange. Anything that is relevant to the analysis of a policy problem can be defined as either a cost or a benefit to human individual preference satisfaction. The trade-off between cost and benefit is purported to indicate the most rational and moral policy choice for the decision maker. Under market conditions, one can assume that private benefits and costs are equal to joint or social benefits and costs and, therefore, that the efficient outcome (a solution that is Pareto optimal—Feldman 1980, Chapter 7) can be found purely through support for private and isolated voluntary trade.

However, when the principle of efficiency enters the realm of public choice, the replacement of fungible private goods with collective goods causes markets to fail and replaces trade in a market with the use of cost-benefit methods to mimic market solutions (Feldman 1980; Tresch 1981). Here it is critical to understand that the choice of cost-benefit methods over cost-effectiveness analysis is the attempt to use efficiency to judge both the ends and the means of a particular policy and not just the least costly means to policy ends chosen on the basis of some variable other than the principle of efficiency (Gillroy forthcoming a).

In the case of environmental policy, the conditions are not "normal" from a market point of view. Price externalities exist that cause social costs and benefits to be both greater than their individual valuations and involuntarily traded to others without their consent or in some cases their knowledge (Page 1973). When applied to public choice under these circumstances, cost-benefit analysis, functioning where these externalities have forced the market to collapse, attempts to compensate for the involuntary nature of transfers and the social nature of excess costs and benefits. The job of the decision maker under these circumstances is to reintegrate any external costs and benefits into a language in which the market can judge efficient outcomes. This process involves translating all intrinsically non-consumptive variables into equivalent monetary costs and/or benefits to achieve a potential Pareto improvement (Feldman 1980, 142–44). In the process, the non-market externalities and their effects will be put into the price system making them subject to voluntary transfer (or allocation by willingness to pay), where social costs and benefits can equal aggregate individual costs and benefits.

The aim is to use government to set the conditions under which the pollution externality can become absorbed in private cost-benefit calculations (v_i), so that decision makers can aggregate these into a social valuation, and recommend an efficient policy outcome. At this point, where social and private costs are one and the same, each actor will only pollute until the personal costs outweigh the benefits from not contaminating the environment.

The premise is that any government decision is, at most, a second-best solution, falling short of the ideal outcomes of market processes (Lipsey and Lancaster 1956; Tresch 1981, 4). The objective is then to allow normal market processes to handle as much of the externality as possible, leaving government intervention at a minimum. In this way, there need be only limited involvement of the non-market sector and efficiency can remain the sole motivating principle of policy making.

As an integrated evaluation technique, cost-benefit methods are justified by the market ideal and recommend government intervention only to the extent that potential Pareto improvements can be found without infringing on the sovereignty of individual preferences. Hence, the role of cost-benefit analysis is to rearrange the sociopolitical structure of the externality problem (that causes the market to fail) in such a way that the public goods under consideration become accessible to private welfare calculations, trade, and personal preference satisfaction. Cost-benefit analysis is, therefore, less a method of analysis and more a systematic and comprehensive decision procedure flowing from the normative task of honoring individual choice while locating efficient welfare improvements (Gillroy 1992 a).

The Strategic Nature of the Polluter's Dilemma

The character of the environment is joint and non-excludable. It is joint because it is not transferable in the economic sense but exists for one and all simultaneously. The provision of clean air for one person does not significantly diminish the clean air left for others to consume. Non-excludability is derived from the fact that environmental quality is not a private good that can be claimed by one person to the exclusion of others. The property of environmental quality is attenuated (i.e. non-exclusive, incompletely specified, and privately unenforceable). Environmental quality exhibits joint nonvoluntary transfer and non-excludable right, which are traditionally described as the two characteristics of a public good (Hardin 1982; Snidal 1979; Feldman 1980, Chapter 6). However, following Hardin (1982), we will use the term *collective good* to characterize provision of goods through the solving of a collective action problem. In this way, the character of the good itself is not under consideration but rather the fact that its provision is considered collective.

Although a variety of game formats can be used to discuss a collective goods problem (Taylor 1987), we will characterize the provision of a collective good as a prisoner's dilemma situation (Hardin 1971; 1982), where individual self-interest causes a joint strategic interaction that creates a nonoptimal collective provision of goods. For example, consider public television as a collective good. People will want to watch the programs but may not want to pay for this joint and non-excludable good when they can watch without paying. However, if all viewers follow this action, no one will pay and the station will fold for lack of funds. In order to persist, the television station will have to encourage a critical mass of individuals to provide funding. If all free-ride, no public good will exist; but if a critical mass (k of N viewers) does support the station, then the others can free-ride without causing the demise of the station. This (k) can be one person or one million people but represents the minimum number of cooperators necessary to ensure the provision of the collective good.

In the case of environmental policy, the quality of the environment represents the collective good under consideration, and pollution of this environment by individuals causes its quality to be lowered. If we hold to the economic efficiency assumptions of self-interest and preference satisfaction and place this economic model into a strategic framework, we can examine pollution within the context of game theory. Following Robert Goodin's example (1976), we shall describe the prisoner's dilemma involved here as the polluter's dilemma.

> Pollution control is a coordination problem for all those who would rather no one (themselves included) pollute to everyone polluting. Independent decision-making is guaranteed to lead to a situation in which everyone (or nearly everyone) pollutes with, of course, some polluting more than others. Joint decision-making is required if actors are to achieve an outcome

> which virtually all prefer, everyone refraining from polluting. (Goodin 1976, 166)

We set up the preferences of individuals in the following way:

Polluter's Dilemma

(best) 1. Unilateral polluting
2. Universal restraint
3. Universal polluting
(worst) 4. Unilateral restraint

It renders a strategic matrix as shown in Figure 1.1. In this matrix, Row (R) and Column (C) have two strategy choices: to cooperate in the provision of environmental quality by not polluting (R1,C1) or to continue to pollute and free-ride on the quality of the environment (R2,C2). The contention is that individuals, considering their own self-interest in isolation, will have a tendency to want environmental quality but will fear that others will free-ride on their cooperation and take advantage of them. The collective result of this individual, self-interested calculation is an equilibrium at (R2,C2), where both actors play their dominant strategy[2] and continue to pollute, free-riding on one another and eroding the quality of the environment. In this strategic interaction, it is assumed that the actors are making economic cost-benefit calculations that result in the non-optimal provision of the collective good.

> Comparing the costs and benefits of unilateral restraint as against universal polluting, there is one very large cost that is common to both: the pollution produced by everyone else. The question is simply whether the individual ought to add his own little bit of pollution to that vast amount that would be in the environment anyway. A man always gains something from polluting (he reduces costs of waste disposal), so there is a prima facie case for the rationality of preferring universal polluting to unilateral restraint. Hence, there is a prima facie case for the Polluter's Dilemma model. (Goodin 1976, 167)

Considered from this standpoint of economic justification and cost-benefit evaluation, it seems that the self-interested, isolated efforts of individuals to satisfy their preferences will result in an aggregate that favors universal polluting. The cost-benefit analyst, attempting to reflect preferences, will be faced with preferences that, because of the strategic reality of interpersonal relations, will result in a non-optimal collective provision of the good.

In order to provide the collective good, it is necessary for a critical mass of individuals to overcome the strategic reality, to comply and not pollute,

		Column Player	
		C1	C2
Row Player	R1	2 2	4 1
	R2	1 4	3 3

Figure 1.1 The Polluter's Dilemma

so that a standard of environmental quality can persist over time. With an N-person society, this requires central coordination.[3]

> There is no quality in human nature, which causes more fatal errors in our conduct, than that which leads us to prefer whatever is present to the distant and remote, and makes us desire objects more according to their situation than their *intrinsic* value. Two neighbors may agree to drain a meadow, which they possess in common; because 'tis easy for them to know each others mind; and each must perceive, that the immediate consequence of his failing in his part, is, the abandoning of the project. But 'tis very difficult, and indeed impossible, that a thousand persons shou'd agree in any such action; it being difficult for them to concert so complicated a design, and still more difficult for them to execute it; while each seeks a pretext to free himself of the trouble and expense, and wou'd lay the whole burden on others. Political society easily remedies both these inconveniences. (Hume 1740, 538 [emphasis added])

The necessity of government intervention to counteract the economic self-interested behavior of individuals and to maintain environmental quality (Goodin 1976, 168), the establishment of Hume's political society where government is responsible for the provision of collective goods through the coordination of individuals, relies on a concept of critical mass that describes the degree of coercion and coordination necessary for the adequate provision of the collective good.

In the public television example, the public good can be provided by k of N individuals. One, six or any number that are willing to pay can see that the good exists at an optimal amount, while the others in N free-ride without affecting the provision of the good. The public good of environmental quality has a different concept of critical mass. It is not possible for one or a said number of individuals to provide a quality environment, nor ought it to be a private responsibility. It requires that a sizable proportion of individuals, k/N, comply in order for the collective good, in this case, to be provided for all who could be considered to have a joint responsibility for its provision.

It is not enough, however, for a proportion of cooperators to be established, for there will be an absolute number of defectors (N – k) who will be able to free-ride at any level of environmental quality. The pertinent critical mass concept, therefore, is the proportion of cooperators to non-cooperators that allows the maximum number of free-riders (that raises total welfare [Goodin 1976, 181]), taking into consideration the number of cooperators necessary to provide environmental quality at a level that allows the biosphere to persist over time.

> The biosphere is characterized by self-renewing, self-cleaning mechanisms. . . . There are, however, quantitative limits to the biosphere's capacity for self-renewal. . . . Anything in excess of these limits will stay in the atmosphere, river or ocean and cannot be removed naturally. (Goodin 1976, 151)

The proportion of cooperators to free-riders will set the critical mass for any level of environmental quality (Q).

$$Q = \frac{k}{N - k}$$

For critical mass maintenance, the state will have to be involved not only in setting standards for environmental quality but also in coercively coordinating a large enough proportion of the population, based on the ratio above, to ensure that those standards can persist as the population increases. It will not be enough to encourage k, through minimal measures, to cooperate because the rest of N may free-ride; as long as (N – k) is large enough, the quality of the environment will be substandard. We contend that it takes more central coercion to coordinate k/N when one must be concerned with (N – k) rather than k of N.

The critical mass concept supports the argument that the state must intervene to establish environmental quality standards and to ensure that the required proportion of individuals in the society cooperate in the provision of the collective good. It supports the contention that solving the polluter's dilemma will require institutions larger than the minimal state in order to establish a critical mass for the collective good of environmental quality. In addition, the status quo point has a non-static nature, for within the polluter's dilemma it results in a condition where everyone continues to pollute at will and the environment deteriorates. In the normal polluter's dilemma, the equilibrium is a status quo where each player can keep his entitlement. In the case of pollution, failure to cooperate results in a state of affairs in which one cannot maintain a status quo level of environmental quality. In the latter situation, the establishment and maintenance of critical mass is more important while the status-quo is environmental degredation and not a safe stasis. In order to overcome the externality of pollution, it will be necessary to have a greater level of central coordination and to take into consideration that the number who free-ride (N − k) may have a noticeable effect on environmental quality.

Finally, the logic of cost-benefit methods depends on the ethical contention that individuals ought to be "willing to pay to have pollution reduced in order to consume clean air" (Bish 1971, 122). But this moral symmetry between polluter and polluted seems not to take the nature of the polluter's dilemma seriously. The ethical asymmetry between polluter and victim must be seriously considered as long as one can ask why they should have to pay someone to stop harming them. As long as the strategic reality among polluters is akin to the polluter's dilemma, where individual action breeds collective harm, then the role of government as representative of the public interest in the provision of collective goods must be assumed to be real. The asymmetry becomes more critical to policy within the context of environmental risk.

Environmental Risk and the Imprisoned Rider

> A new type of environmental problem is emerging which differs in nature from the more familiar pollution and resource depletion problems. This type of problem, which may be called environmental risk, has rapidly increased in importance over the last few decades and may indeed become the dominant type of environmental problem. (Page 1978, 207)

It is the strategic nature of environmental risk, its moral ramifications, and the political institutions necessary to manage these three variables that are of primary concern in this section of the chapter. We will define the parameters of the public policy choice involving environmental risk, which in turn will inform the construction of a justification and evaluation system to handle these critical cases.

The Unique Character of Environmental Risk

The seminal article that distinguished the character of environmental risk from traditional pollution cases was written by Talbot Page (1978) and published in *Environmental Law Quarterly*. Page listed nine distinct characteristics of environmental risk, four having to do with its uncertainty and five others that concern its institutional management.

The first distinguishing characteristic of environmental risk is related to a technical *knowledge of mechanism*. In traditional pollution cases (litter, detergent in water, smoke and particulates in the air) the generation and transmission of the pollutants are understood by science, as are the appropriate responses to these types of pollution. In many such cases, the presence of pollutants is detectable to the naked eye of the technically unsophisticated observer.

The mechanism by which risk is created and transmitted to people and the environment is not well understood, which makes decisions about proper means and amount of control, or what defines control, a more difficult proposition. With environmental risk cases (toxins in groundwater; generation of nuclear power and disposal of its waste; production, storage and use of synthetic chemicals and pesticides; and genetic research and experimentation), suspected toxins, alone or in reaction with others, could produce carcinogenic, teratogenic or mutagenic results that may be unnoticeable even to a trained eye. Overall, "the mechanisms of generation, transmission, and response are understood so poorly that any management of these problems is truly decision making under pervasive uncertainty" (Page 1978, 209).

The second characteristic of environmental risk is its *potential for catastrophic costs.* In cases of traditional pollution, such as suspended solids or eutrophicants, scientific understanding of the mechanism at work makes it possible for decision makers to judge both the degree of risk involved and the potential costs of worst-case scenarios. The pervasive uncertainty related to environmental risk makes costs unknown and worst case hard to define. It is assumed, however, that the costs are much higher and less incremental than in cases of traditional pollutants. The synthetic nature and large number of prospective toxins, for example, and the theoretically potential synergistic effects of mixing them, as well as the fact that smaller doses of environmental risk contaminants pack higher concentrations of potency, lead one to believe that the effects of environmental risk could be catastrophic. "But what is not known about mechanisms precludes specification of just how catastrophic and likely the costs might be" (Page 1978, 209). The obvious and incremental nature of most traditional pollutants that are amenable to cleanup must be juxtaposed with the clandestine and lump-sum nature of environmental risks that may result in catastrophic costs in terms of human and environmental persistence. This is a juxtaposition that must be taken into account in the formulation of policy on environmental risk.

The third feature of environmental risk cases is their *relatively modest benefits.* Although some might feel that the benefits of nuclear power, synthetic chemicals, pesticides or recombinant DNA are not modest at all, the positive aspects pale in comparison to the potential long-term harm to human and environmental life. A large-scale release of radioactive contaminants, a toxic chemical spill, or an atmospheric leak of toxins (Bhopal, India), would cause one to think that "there appears to be at some level a strong asymmetry between potential costs and benefits"

(Page 1978, 210). Any of these occurrences could lead to widespread death and long-term mutation in the genetic code of humans and other life-forms.

The final characteristic related to the uncertainty of environmental risk is its *low subjective probability.* "Due to the fragmentary knowledge of mechanism, the likelihood of the catastrophic hypothesis cannot be determined objectively, but must be assessed subjectively, based upon whatever knowledge is available" (Page 1978, 210). The subjective probability connected with environmental risk is normally low, which introduces a second asymmetry.

> Low subjective probability of the catastrophic outcome . . . tends to counterbalance the asymmetry of potential high costs and modest benefits. . . . Whether the greater likelihood of the favorable outcome compensates for its smaller relative size is a fundamental question of environmental risk management. (Page 1978, 211)

The reality of low subjective probabilities of infinitely harmful costs labels environmental risk as a *zero-infinity problem* (Page 1987).

In addition to these four characteristics related to the uncertainty of environmental risk, Page defines five characteristics that affect the institutional management of environmental risk. The first two have to do with the generation of costs and benefits in environmental risk cases—specifically, the *external transfer of benefits* and the *external transfer of costs.*

With environmental risk, the small benefits of accepting the gamble are allocated within markets (the greater profits of firms, greater consumer satisfaction) involved with the development, production and marketing of technological innovation within the price system. On the other hand, "the adverse effects [costs] of environmental risk gambles usually are transferred directly through the environment rather than through the market" (Page 1978, 212).

The externality costs of environmental risk, like traditional pollution cases, are collective in nature and involuntarily transferred. The difference is that the gamble, inflicted on the general population by the few who reap benefits from environmental risk, is a zero-infinity dilemma. Risk externalities in zero-infinity dilemmas have much higher potential costs borne by a population that collectively has not accepted the risk voluntarily. *The collective nature of the risk* is the third unique feature of environmental risk management.

The central concern is that the diffusive effects of environmental risk pose severe questions about traditional institutional ways of protecting individuals from risk gambles in public policy. Such protective mechanisms as insurance, liability law or other compensatory systems are inadequate to the task of protecting, ex ante, the millions of people potentially affected by environmental risk (Page 1978, 213).

Page condemns as inappropriate for environmental risk policy the policy tools related to all market conceptions of "socially better." Markets, seeking potential Pareto improvements, can only promise ex post compensation, which is not protection ex ante but only an attempt to pick up the pieces after the fact. Because of the reality of environmental risk, its potentially high costs, and its collective nature it is necessary that individuals be protected before the fact and that whatever risk is accepted is distributed fairly, especially among those who reap the benefits. These concerns can only be addressed on a collective level and therefore call for political institutions and a sense of ex ante political action to assess risk and distribute it equitably within a population. The need for political institutions and a non-efficiency-based set of values to inform the definition of "socially better" and shape policy ex ante is further supported by the final two management characteristics of environmental risk: *latency* and *irreversibility*.

> The extended delay between the initiation of a hazard or exposure to it, and the manifestation of its effects . . . is sufficiently long and the risk sufficiently diffuse so that the risk is borne involuntarily, if not unknowingly. . . . Long latencies [also] increase the likelihood that the potential effects of environmental risks will be masked by other factors. (Page 1978, 213)

The collective risk is such that one may be affected by a toxin and not know it until ten or twenty years later when health effects surface. Latency may even stretch over many generations, who, as the result of losing the environmental-risk gamble, are faced with changes in genetic coding that cause mutation over time.

In addition to the factor of time and its effect on information and on individual capacity to perceive harm, the effects of environmental risk are often irreversible. They are not privately irreversible like consuming a hamburger but collectively irreversible. One individual or firm, who gambles and loses the zero-infinity risk, will impose an irreversible harm on the community that may not become evident until years or generations later.

Page is correct to conclude that the collective nature of these characteristics has "profound ethical and institutional implications. . . . They raise questions concerning fair distributions of risk over time and how institutions can be designed to anticipate adverse effects, rather than merely to react to existing known effects"(Page 1978, 214).

These ethical and political questions are about moral principle, systems of valuation and the proper ethical justification of political institutions.

They speak to the reality of environmental risk as a species of collective action problem requiring both a distinct set of values that can shape normative argument and institutions that relate directly to the character and requirements of human beings and their natural environment independent of efficiency considerations.

To the characteristics of environmental risk itemized by Page, we would add one more: its *strategic nature* (Gillroy forthcoming b). The strategic reality of the zero-infinity gamble is in some ways similar and in other ways dissimilar to other environmental collective action problems but is completely distinct from markets and their assumptions.

Environmental risk is a collective action problem. In order to anticipate and distribute the risk fairly, central guidance will have to coordinate actors, sometimes against their subjective welfare preferences, to establish a critical mass that can ensure that environmental quality is achieved and maintained over time. This being the case, we can use the polluter's dilemma, with certain adjustments, as the basic model of the strategic reality of environmental risk situations.

A unique characteristic of the environmental risk dilemma is the goal of coordination. The responsibility of the central authority is not so much to provide a collective good, which may already be assumed to exist, but to prevent a public bad from causing collective and irreversible harm. The provision of a collective good and the prevention of a public bad can be distinguished strategically by their unique concepts of critical mass.

In the case of provision of the collective good, it is the goal of the coordinating authority to get k or k/(N – k) players to override their subjective self-interest and cooperate. Although the number who defect will lower environmental quality, with these critical mass thresholds it is possible for some to free-ride without hurting the provision of the good. In the case of preventing a public bad, free-riders cannot be so graciously tolerated, and the coordinating authority must use N – k as the sole proper critical mass concept, not just as part of the proportion of cooperators to free-riders. The imperative of political institutions is to keep N – k to a morally justified minimum.[4]

Environmental risk is such that small doses of highly toxic material can have catastrophic effects on a large number of actors. It would not require many free-riders (defectors from the critical mass that provides the collective good) for the public bad to render widespread harm. The low threshold for free-riding (N – k) on the critical mass that is cooperating to prevent the public bad makes it harder to achieve the cooperative solution. When the goal is to minimize the number of noncooperators, it will in everyone's interest to be part of the critical mass if all others (or nearly all others) are. Those who cooperate must expect that central institutions have established and justified standards, and have the power and resources to apply them and detect defectors from cooperation swiftly to prevent long-term harm to the community. One will cooperate only if the definition of acceptable risk is clearly defined and if the system of justice that distributes it, has proper moral justification that is understood to set the expectations with which everyone joins the critical mass.

Another distinctive characteristic of the environmental risk dilemma is that it is a second-stage prisoner's dilemma. One must assume that the polluter's dilemma is solved in that a standard for environmental quality has been set by the central authorities and this standard is protected by the cooperation of the vast majority of actors in the society, based on a critical mass of $k/(N - k)$. The task of the authority is then to impose stricter cooperative requirements, minimizing $(N - k)$ in cases of environmental risk and preventing the breakdown of the collectively optimal solution (R1,C1) (See Figure 1.1).

A final consideration in the strategic reality of environmental risk involves the strategy choices of the players. The normal polluter's dilemma gives each player two strategy choices: to cooperate in the critical mass providing environmental quality or to defect and pollute. In most instances, environmental risk cases involve one collective actor (the firm) that develops a technology or scientific process that presents the risk and the rest of the population, who share the collective dissemination of costs from any risk. The reality is that only one of these parties has the option to impose the risk, while the other, without collective representation, can only accept the risk's eventual outcome. This is especially true of environmental risk dilemmas where risk is not evident to one's senses. One can live in a house full of radon or drink water laced with toxic chemicals and not know it. In addition to the long latency period before physical ailments become evident, individual citizen-players may believe they exist in a low-risk environment when they do not. Without the collective energies and resources of political institutions set up to support the citizen's full agency, the other player (firm) can use its dominant strategy without fear of detection nor retaliation.

The individuals who make up the population are therefore faced with a risk that is potentially catastrophic and irreversible, has long latency periods that conceal it from their knowledge or perception, and that affects them collectively. Therefore, even an astute individual citizen acting alone has little power against a firm who wants to market a new risky technology. If we look at the prisoner's dilemma game matrix and extend the efficiency assumptions about the individual to the case of strategic interaction and environmental risk (see Figure 1.1), we can see that if one assumes that no background institutions exist the equilibrium will not be at (R2,C2) but at (R1,C2).

Individuals have no choice but to cooperate. They have no coercive threat power (i.e. defection and counter-imposition of environmental risk) and require the collective good of environmental quality to continue to live. Therefore, row (the individual) has no alternative but to choose his lone strategy R1. Column (the firm facing unorganized individuals), receiving no disincentive from the other player, has the prima facie inclination to defect in order to maximize its internal benefits. Also to column's advantage is the reality that it can impose the risk with stealth, that is, without the direct knowledge of the other player, who has neither the power, skill or resources to monitor and find an environmental risk violation nor the capacity to do anything about it if it were detected. The firm will therefore be able to serve its self-interest by choosing strategy C2, without fear of detection or reaction from the other player.

In cell R1,C2, the individual is exploited by the firm, receiving the lowest payoff while the firm gets the best. We will contend that in cases of environmental risk, the lone individual is strategically an *imprisoned rider* (Gillroy forthcoming b). We will further contend that central political representation, based on the individual's interests as a member of the collective that exists to protect and define the public interest and to prevent the exploitation of isolated individual citizens, is a necessary prerequisite to freeing the imprisoned rider. We will maintain that only central authorities have enough organization, power and resources to coordinate (N − k) actors and anticipate and stop any defections that could render harm to all. Only with a morally justified central authority to represent individuals as citizens will it be possible to manage risk and distribute it fairly within and between generations in such a way that R1,C1 persists as a point of equilibrium and all are protected from exploitation in R1,C2. Active state institutions are necessary in order for the row player to have a threat strategy against column, in order for individuals to have the capacity to protect themselves from being exploited and in order to reestablish a symmetry of strategic choice between the players. Political institutions give individuals the power to preemptively require the cooperation of firms that, in order to be efficient, may not have subjective interests in that direction.

Harm to Intrinsic Value: Beyond Monetary Cost

The cost-benefit approach to policy decision making in cases of environmental risk, like its approach to any and all areas of policy, is preoccupied with what it is designed to prevent and what it is best capable of handling: instrumental monetary cost to the economy. The money that will be lost or gained and the opportunity cost of capital expended in the achievement of other ends are the permanent and sole concerns of one using cost-benefit evaluation, where a benefit is defined in terms of a dollar cost saved. This overriding interest in funds and their allocation is not solely a concern with the efficient allocation of capital but is deeply tied to a moral intuition that maximization of wealth in trade also provides for the good of the community.

In cases of environmental risk, there are indeed monetary costs to be considered, both real and in terms of economic opportunity. It costs real dollars to clean up a toxic dump, to properly dispose of waste materials, or to build safety into a nuclear reactor. There are opportunity costs when a firm must test extensively and hold back a new chemical or genetic technology, sometimes only to have it prohibited from sale. Are these the sole costs of environmental risk? If not, are they the most important costs? If no again, what are the central and most critical negative effects of environmental risk?

We will answer all of these questions through the examination of the last one. What are the critical costs of environmental risk? Costs are to be found in the effects that can be expected in the "worst case" scenarios of environmental risk. Although scenarios differ and are, as we have seen, uncertain, it is possible to describe what is at stake when these zero-infinity risks inflict their maximum costs. We will assert that the point of departure for cost analysis in environmental risk is the capacity of humans and ecosystems to persist and reach their full potential within their social and/or natural environments. It is the "integrity" of human beings and the ecosystems in which they exist, the intrinsic values of the moral capacities of humankind and the functional capacities of the natural environment that are at stake in environmental risk gambles.

The natural environment is an intricate set of interrelationships and elemental cycles that are put under pressure by environmental risk costs. Due to the synthetic nature of the technologies under consideration and their irreversibility and mutagenic effects, environmental risk contaminants are potentially more dangerous to ecosystem persistence and natural selection than traditional pollution problems that, to some degree, could be accommodated by the natural cleaning cycles of the environment. The breakdown of a number of critical subsystems could potentially cause a dysfunction of the entire natural environment, a process of degeneration that may be irreversible once it has started.

Humans, exposed to the effects of environmental risk contaminants, are dependent on the natural environment (as the medium of transmission) for their ability to think and process information for decision making. Here the assault of toxins, radioactive material and new strains of synthetic disease will directly affect the air they breathe, the water they drink, the food they eat, and the land on which they live. The capacity to think for oneself and to have a personality or self are the central components of each and every human being and are in jeopary in the environmental risk gamble. Without the capacity for self, the individual can have no independent personality, no conscious thought or purposeful action, nor any possibility for preferences, let alone a rational preference order that is, in any real sense, one's own.

To use the word *cost* in relation to what is at stake is, at best, inadequate and perhaps a logical category mistake (Sagoff 1981). The human capacity for agency (integrity) and the natural capacity of systems of natural selection to persist over time are concerns that ought to be put into terms that reflect their intrinsic value. Cost makes sense only in relation to variables that have a monetary price and can be traded as fungible goods and instrumental means for the end of wealth maximization (although cost is also used as a synonym for uncompensated economic loss [Pigou 1932]). Harm can be related to the incapacitation of that which has intrinsic value. Harm, for our purposes here, will be distinguished from cost as that negative result of environmental risk that affects the intrinsic capacities of man and natural ecosystems. In this definition, harm carries a moral connotation that relates to the damage of the individual's essential interests. Harm also gives an asymmetric weight to preventing a violation as opposed to providing a benefit. It is no longer the case that costs and benefits are symmetric concepts; prevention of harm, due to its effect on essential value, has priority over provision of a benefit in welfare terms. Harm is established in opposition to non-benefit, which is the economic definition of cost (Feinberg 1973, 29–31).

Contrary to our definition of harm, the cost-benefit criterion assumes that avoiding a harm of a certain magnitude is just as desirable as producing a benefit of the same size. We maintain that this symmetry is not consistent with the gamble in environmental risk. Harm to integrity is the critical consideration; it is separate and more important than cost to the economy or any individual in the market and is therefore not subsumed in the efficient maximization of wealth nor in instrumental preferences for welfare improvements. Capacity is prior to preference in that without a self preferences cannot be created, and without conscious capacity there is no defined self (Fried 1977, 185–88).

The effects of harm associated with environmental risk will not be prevented or remedied by the encouragement of subjective wealth maximization, for this behavior is what causes the strategic reality of the polluter's dilemma and the imposition of risk by the few on the many. Nor will environmental risk be rendered harmless by the noninterference of government, for it is the lack of an active collective agent for the public interest that allows the environmental risk prisoner's dilemma to produce the phenomenon of the imprisoned rider. Nor will cost-benefit evaluation be adequate to define "socially better," because it is not instrumental monetary cost that is at stake but intrinsic harm to humans and their environment. An economist cannot accommodate such an asymmetric concept as "intrinsic" cost but can only treat it as a non-benefit.

The proposition is that the capacity for human thought and action and the capacity of the natural environment to persist are at stake in environmental risk policy. These are basic, necessary, and comprehensive concepts of intrinsic value because they concern the internal integrity of living and evolving organic, and in the case of the individual, moral capacities. Respecting essential freedom within the context of environmental risk requires that the central authority anticipate free-riders and regulate their behavior so that the capacities of its citizens are protected and empowered ex ante, before economic externalities can harm them. The need for regulation of economic market behavior and an active state that can protect and empower the capacities of its citizens to think and act for themselves requires a justification scheme that moves beyond preference, consent and "willingness to pay" as the bases for policy choice.

Respect for Intrinsic Value

Moving beyond superficial considerations to the essential value of man and nature will reorient or redefine those concepts associated with public policy decision. Specifically, a theory of integrity at the foundation of policy decision making will introduce a new language of right and duty into the evaluation and justification of policy. From this vantage point, the problems of environmental risk involve a lack of respect for the integrity of human agency, for the potential expression of full capacity, by those who would exploit the imprisoned rider and impose risk to further their economic interests.

The consideration of an agent's capacity based on a principle of integrity grants a specific character to right and duty. It is not just an interest of the individual to be a capable agent but a right that invokes a private and a collective duty to create and support those political institutions that regulate the costs, benefits, harms and potential exploitations of the economic behavior of any agent toward the rest of the population. A right is defined as a claim one actor makes for himself as an individual or as part of the political community to that which empowers or protects his internal capacity for agency. Each right that is claimed involves a corresponding duty (on an individual or collective level, depending on where potential violation of the right is centered) that one provide for one's own integrity and the agency of others in the political community. Here freedom involves obligation to ourselves and to others as intrinsically valuable entities in an environment that supports us, while it exists with a purpose, potential, and therefore an intrinsic value of its own.

> We can be truly free to pursue our ends only if we act out of obligation, the seeming antithesis of freedom . . . to be free is not simply to follow our ever-changing wants wherever they might lead . . . the highest purpose of human reason is to evolve a comprehensive understanding of mankind's place in the universe, not merely to serve as a detector of consistency and causality and thus as an instrument for morally blind desire. (Tribe 1974, 1326–27)

The policy maker ceases to be interested merely in aggregating the diverse subjective ends of distinct individuals to maximize their wealth or happiness but is concerned with respecting individuals as agents with the capacity to be of moral value in and of themselves. The concept of duty replaces the reliance on the consequences of preference aggregation for the efficiency of social policy. The moral requirements of such a duty that would empower the state to regulate in anticipation of free-riding must be established and justified if the state is not to classify any and all action as that which is compatible with its collective duties. Coercion must be mutual and morally grounded.

In order for the state to protect, ex ante, the essential choices of individuals, it must regulate that which interferes with the full potential growth and use of agency as intrinsic value: those material conditions that retard the integrity of its agent-citizens. The state is responsible for the provision of information but also for the distribution of the material entitlement, the opportunities and the level of environmental quality that directly affect the ability of individuals to use their freedom and autonomy in making choices and acting in the polity. These material conditions form a baseline that externally defines that without which individual integrity (essential internal freedom) cannot flourish. Basic entitlement and opportunity provide the external conditions in which the internal capacity of the individual can find its full range and potential expression, while the level of environmental quality protects one's rationality from the ravages of toxins. Overall, equal capacity is fostered and protected by equal distribution at the baseline. The strategic nature of environmental risk makes collective representation in the provision of this baseline a necessity.

Environmental risk decisions, therefore, ought to be made on the basis of how each alternative will affect the baseline factors, in order for the decision to reflect the intrinsic values at stake. Life would not be a quantity of time but a quality of evolving or persisting mental capacity. The baseline of goods, opportunities and environmental quality necessary to essential persistence of any person's integrity as a human being and the functional integrity of the environment combine to offer the essential foundation for environmental policy. Any trade-offs should be considered in the light of how one type of integrity is affected by the other. One assumes that the environment will contain risk. The question is, how does this risk affect human moral integrity and environmental functional integrity? Risks imposed ought to be judged on the basis of the effect of the decision on the rights and duties implied by a respect for integrity and not on the basis of cost as the primary factor in the judgment of the ends and means of environmental policy. But what does this mean?

In terms of environmental risk as a collective action problem, individuals have a private right to respect for their integrity and a public duty to respect the integrity of others. They will be assumed to have a public right to their baseline quantities of public and private goods, opportunity, and environmental quality and a private duty to participate in and support the collective institutions that protect and empower their capacity and the integrity of others, without which compliance in collective action can neither be expected nor morally required. Environmental quality and environmental risk should be collectively regulated to the advantage of each person's full integrity and the functional integrity of the natural environment. In cases where these conflict, trade-offs should be equitable. Intrinsic value can be exchanged only for another intrinsic value (e.g. basic food for one less woodland, the sacrifice of higher crop yield for the protection of a unique ecosystem).

By this definition, good environmental quality is a condition that we have a public right to expect and a collective duty to provide. We may rightfully expect that we and the environment will be protected from the externalities of those over whom we have no personal control, especially when they could affect our essential freedom or the persistence of natural systems for instrumental efficiency reasons alone. On this basis, we are not privately expected to pay for what we have a right to as individual members of the public at large. Our public duty to respect others coincides with our private duty to participate in the political process that provides collective compliance and this, in turn, defines what is required of each individual in a collective situation in order to insure the freedom and integrity of all on an individual level.

It is no longer important for policy makers to decide what an individual is willing to pay for. It is critical, however, that decision makers understand what each person has a private right to expect from them and that they have a public duty to provide for their fellow citizens and the natural environment. This sets the expectations of evaluator and citizen alike, the range in which cooperation can be expected and how this cooperation will be defined and any conflicts resolved.

Environmental Risk and Environmental Values

The critical foundation for economic cost-benefit analysis consists of the internal assumptions and ethical prejudices that construct the way in which the economist views the world, which variables count in policy considerations and by how much, and, ultimately, how far the market can go in pursuing efficiency.

Is the efficient satisfaction of individual preference to produce welfare the proper means and ends for environmental policy, regardless of how well this might be done within cost-benefit evaluation? Or is it necessary to move beyond preference and efficiency in order to establish an adequate definition of "socially better" because of the strategic and moral asymmetries involved with environmental risk policy?

The strategic reality of environmental policy is such that it requires a moral asymmetry to be taken into consideration in the evaluation and justification of public decisions. This moral asymmetry has two dimensions. First, we must distinguish the party to the strategic interaction that inflicts the uncompensated loss from the party that suffers costs from externality-producing behavior. That is the isolated free-rider, whose economic decision making imposes the cost on the collective (all others), which is not always empowered to adopt defensive strategies. Second (especially relevant in environmental risk cases), a dichotomy should be recognized between essential harm and nonessential costs, between intrinsic and instrumental value. In the provision of a collective good like environmental quality, both of these dimensions entail policy considerations.

The first dimension requires that the policy maker be able to segregate those who inflict the loss and hold them responsible for compensating those they inflict it upon, who have a proper expectation that such compensation is forthcoming. A sound environmental policy demands that the evaluator define a collective interest as distinct from the sum of individual interests. It must be assumed that the collective good will often conflict with individual choice and preference, and therefore political institutions must be constructed, and terms of conflict resolution established, so that coordination can be established that is fair to both individuals and their collective interests.

Within the logic of market efficiency and its use of cost-benefit methods, the state may absorb transaction costs that inhibit market exchange but cannot regulate that exchange once it has the proper foundation and has begun. All values in the calculation of proper means and ends are symmetric, depending on the disposition of property rights where who is assigned the rights is considered morally arbitrary. The only state action that is called for in this case is that which assigns these rights and provides the expectations and contract enforcement. No independent collective interest exists in this analysis, only the aggregate of individual economic choices that allocate by each person's willingness and/or ability to pay for environmental quality.

A limited level of state involvement in the setting of environmental policy is troublesome within the context of traditional pollution problems and the polluter's dilemma, but when one moves from the regulation of traditional pollution problems to environmental risk this role seems even more strained. In the case of environmental risk, with its long latency periods, irreversibility and the moral implications of the imprisoned rider, the collective "good" must be able to take precedence over individual wants or the unfettered functioning of market processes. It is more important than ever to consider the proper distribution of burdens and risk, to have a standard for the equal consideration of all, now and in the future, and to see that those who might be harmed if it were not for the protection of anticipatory institutions are treated fairly. Considering what is at stake, is the value system of cost-benefit methods, with their emphasis on instrumental calculations of incremental policy, adequate to the task of regulating environmental risk?

> Incrementalism is especially inappropriate to environmental policy, however, because we have no idea of how long to wait before judging the consequences of one incremental change. Without understanding the workings of the ecology, we have no idea how long a policy change must percolate through natural processes before being reflected in some concrete change in the environment. (Goodin 1976, 153)

We have contended that what is at stake in environmental risk situations is the integrity of individuals and ecosystems. How can we define our basic policy values so that integrity is protected and the moral asymmetries are not ignored? The answer lies in the distinction between intrinsic and instrumental value. By defining integrity as an internal capacity, we will generate a set of values concerned with protection, conservation, and fair distribution and define in this way the responsibility of the active state. This will require us to move beyond subjective preference, choice, and efficiency as the sole means and ends of policy, and therefore to move past reliance on cost-benefit methods to measure the proper means and ends of public choice.

Cost-benefit analysis is based on the market justification system and promotes efficient wealth maximization. To be part of market analysis, all variables must be only of instrumental value to the maximization of wealth based on the efficient satisfaction of preferences, which is the sole essential value of the market. The wants of individuals and the satisfaction of these, in welfare terms, are the central values generated from the principle of efficiency, which when applied to environmental policy by cost-benefit analysis will cause the environment to be valued to the degree it contributes to the maximization of net immediate monetary benefit over expected economic cost. The natural environment and the human agents that inhabit it, and to some degree control its destiny, will have no independent weight in the competition for resources, wealth and preference satisfaction.

> Policy analysts typically operate within a social, political and intellectual tradition that regards the satisfaction of individual human wants as the only defensible measure of the good, a tradition that perceives the only legitimate task of reason to be that of consistently identifying and then serving individual appetite, preference, or desire. This tradition is echoed as well in environmental legislation which protects nature not for its own sake but in order to preserve its potential value for man. (Tribe 1974, 1325)

Cost-benefit evaluation will assume that the environment, like all other considerations, has value only as an instrument for man's want-directed nature (Bagley 1961, 149–50). Consequently, if an artificial technology serves man as well as a natural competitor, then it is also a morally acceptable substitute. As long as the level of satisfaction is the same with both, they are considered fungible goods subject to trade. The quest for the potential Pareto improvement and the reliance on preferences exogenous to the mechanism of the cost-benefit evaluation, reduces all ethical considerations to those in search of welfare-maximizing options where the environment is just another good that can be traded off against others until the costs to man outweigh the benefits.

> By treating individual human desire as the ultimate frame of reference, and by assuming that human goals and ends must be taken as externally "given" rather than generated by reason, environmental policy makes a value judgement of enormous significance . . . any claim for the continued existence of . . . wilderness . . . or . . . species must rest on the identification of human wants. . . . As our capacity increases to satisfy these needs and wants artificially, the claim becomes tenuous indeed. (Tribe 1974, 1326)

In order to include a variable in a cost-benefit analysis one must first be able to quantify it in monetary terms. Cost-benefit evaluation is a quantitative technique for computation of the ends and means of policy. Like all quantitative decision-making techniques, cost-benefit is "likely to bias conclusions in the direction of the considerations [it] can most readily incorporate" (Tribe 1972, 97). Those costs and benefits that normally appear in monetary terms will therefore be more "readily incorporated" than qualitative variables like human health, natural ecosystem damage, moral good or right, or aesthetic considerations that are not usually analyzed in terms of money. Everything that finds its way into the cost-benefit trade-off will nonetheless be considered as a priced consumer good. No discontinuities or incommensurabilities between goods will be tolerated by the cost-benefit criterion. All considerations that are part of the analysis must be fungible and placed in tradable currency.

> The objectivist tendency to deny discontinuities by reducing everything to varying levels of smoothly interchangeable attributes (inevitably related to human satisfaction) seriously limits the capacity of the policy sciences to accommodate certain kinds of value . . . values that are intrinsically incommensurable with human satisfaction . . . values with inherently global, holistic or structural features that cannot be reduced to any finite listing . . . of independent attributes . . . [and] values that have an on-off character like . . . integrity, . . . ecological balance . . . or wilderness. Whenever values in any of these major categories appear to be involved, . . . one should recognize that the techniques of policy analysis . . . will tend either to filter them out of the investigation altogether or to treat them in ways inconsistent with their special character. (Tribe 1972, 96–97)

These considerations bring us to a point where we must ask whether the intrinsic value of people and their natural environment, as these relate to moral good or right, ought to count as a unique policy consideration. Should policy analysis be responsible for the protection of humans and their environment as an end in itself? We may want to retain the prerogative to reject a policy that is inefficient, but we might also want a policy justification and evaluation system that is based on a set of values by which we can judge policy on other grounds related to the intrinsic value of humans and the environment.

> Policies may be condemned . . . because they threaten parts of the natural environment. Policies . . . that involve the destruction of wilderness may be condemned because they destroy what has intrinsic value . . . things which are valued in themselves and independently of uses to which they might be put. (Elliot and Gare 1983, ix)

Overall, one might say that for those policy decisions that involve human interaction with the natural environment, cost-benefit procedures of evaluation ought not be the primary concern of policy makers because those procedures do not take into consideration the special character of what is at stake. Based as it is on market justifications, this would require us to doubt the ethical validity of cost-benefit analysis as well as the instrumental principle of efficiency for judging the means and ends of policy. Perhaps it

> embodies an unacceptable premise, namely that the question to be answered is "What is efficient for society?" rather than "What is good for

> society?" By all means use the most cost-effective way to achieve the end once the end has been determined, but do not use the cost-benefit analysis to determine the end. (Ashby 1978, 56)

The determination of ends ought to take the intrinsic value of man and nature as is and set standards for externality control and the management of involuntary transfers on the basis of a set of principles that can handle the unique quality of what is at stake in environmental policy, especially concerning environmental risk, without devaluing it or ignoring its special character.

Political theory can be used to render from the principle of protecting and empowering the integrity (intrinsic value) of humanity and its natural environment a set of "environmental values" that entail an active state with adequate terms of conflict resolution that would be responsible for the coordination and provision of the collective good of environmental quality. With these values in place at the foundation of policy evaluation, the case of environmental risk, even with its requirements for stricter compliance made necessary by the concern for minimizing and anticipating free riders (N − k), may be more adequately debated within the proper context of a stronger system of moral prerequisites.

With the strategic reality of the polluter's dilemma as our point of departure, we must, in defining a public interest based on environmental values, consider the wants of individuals as private, isolated consumers to be the point of strategic stalemate. We must look to individuals as citizens with collective interests as the proper source of direction for the political system. The political institutions, by setting the terms of coordination and by defining the collective good, "define the manner in which citizens are related to each other . . . political institutions constitute the citizenry in the sense of giving it an organized existence" (Elkin 1985, 260–61). In this way the general good, or the collective welfare, takes on an independent existence that is reflected in citizen preferences (Sagoff 1983, 2–7), which are not means to subjective consumer ends but represent concern for the persistence and realization of his collective identity as part of a political community.

The logic of political rationality is distinct from its economic counterpart in that a political policy maker is interested in how the individual participates in the goals of the collective. After all, "political rationality is not concerned with scarcity in the sense of something being subject to consumption" (Elkin 1985, 268) but with individual and collective responsibility to the public good and the terms of coordination necessary to its provision.

In the case of environmental quality, instead of economic efficiency there is *responsibility*, which can be partially defined as political institutions taking the attenuated property rights of a clean ecosystem in trust for their citizens. A joint and non-excludable good ought to have joint and non-excludable rights and a collective entity to oversee its persistence and provision over time.

Responsibility can be further defined in terms of equality and fairness within and between generations. Political institutions must be concerned with the persistence of its citizenry over time and with the equal participation of all in collective decisions (in person or by representative). Because of the existence of involuntary transfers, it is necessary that the political system coordinate fairly, that is, that it not allow one person or subset of citizens to bear an asymmetrically large burden in the provision of public goods. Fairness also includes attention to the distribution of goods and services, in addition to their reallocation by policy. It is important that all citizens participate justly in the terms of political coordination, no matter where in the community they place themselves, or where in time they come along. There will be different definitions of fairness and equality, but any definition must be justified by a systematic set of environmental values that define acceptable asymmetry and determine (1) how and to what extent future generations will participate in the decisions of today, (2) what degree of distribution is fair and just to all citizens, and (3) how the good of the collective is deciphered and promoted.

The principle honoring the intrinsic value of humanity and the environment will produce a set of environmental values. Environmental values in turn will be concerned with the ex ante and/or ex post effects of policy on individuals and with the effort to reconcile the rights of individuals with the good of society. Whatever the basis of interest (utility, resources, opportunity, individuals as ends), environmental values as a foundation for policy decision making will use politics to change the reality of personal interaction from what it is strategically to what it ought to be morally.

Overall, the environmental values will define the intensity of an active government and the particulars of individual and state responsibility based on the degree of cooperative compliance necessary for the persistence of human and natural integrity. In any case, the policy approach will be concerned with establishing a system of decision making that can describe a non-market ideal to solve collective action and collective goods problems responsibly by actively coordinating individual action through central regulation. In addition to empowering the state, the environmental values will restrict and define what is a proper case for state intervention and what is not.

Critics would say that without efficiency and preference, policy can have no objective ethical content, but we will contend that this is not true.

One advantage of the political move toward environmental values is that it challenges one to devise an evaluation system that is a real guide for policy and that reflects the strategic and moral character of collective goods dilemmas.

The specific type of collective action problem involved in environmental risk requires that the intrinsic value of human agency and ecological function be recognized and that the basis of their moral validity shifts from the consequences of policy to the duties and rights that policy protects and empowers. By replacing conventional and relative definitions of fairness and equality with definitions that relate to the intrinsic, universal and necessary values at stake, we can get at the unique character of environmental risk and justify a wider role for government and a more fitting definition of "socially better."

In order to empower and protect human and natural integrity, achieve compliance in the collective coordination of individual action, and ensure that the intrinsic value of the environment is part of policy, it will be necessary for the policy maker to have a strong set of environmental values underlying policy decisions, values that justify case evaluations and recommendations. Market values will continue to rely on ex post legal compensation (liability law) as an acceptable tool to take the moral asymmetry into proper consideration. Environmental values will function ex ante, and not depend on liability law or the ex post transfer of funds to compensate individuals who have had their essential intrinsic value, their full agency, harmed by the acts of other agents. The moral system on which institutional evaluation of policy is based must be strong enough to effect decisions based on duty before harm is transmitted by risk, through the environment, to citizens. The rights and duties involved in the protection of potentially imprisoned riders require that policy makers be responsible in regard to the integrity of their citizens and distribute the baseline fairly and equally across the generations, holding power in trust for all real and potential agents.

The essential distinction pertaining to the adoption of environmental values as the basis for policy decision making is the recognition of an asymmetry between instrumental and intrinsic value. One must be able to give asymmetric weight to the harm that might befall an essential capacity as against the monetary costs or benefits of any specific decision. The fully active state ought to have the power to anticipate free-riding and penalize it, to redistribute wealth and opportunity, and to provide the compliance necessary to maintain environmental quality required to persist in the light of environmental risk. The rights of individuals and the duties of government toward individual agency are such that regulatory responsibility is paramount for the persistence of humankind's essential nature. The burden of proof in the imposition of risk shifts from those who want regulation to those who would support economic behavior that produces a public bad. The point of departure for the policy maker is not that the economy should function without interruption but that individual agency and the functional integrity of the environment ought to be fostered and protected while economic behavior is regulated, unless it can be proved that the imposition of risk through the economy is necessary to the quest for full and equal access to individual and/or ecosystem integrity. This cannot be done without a shift in the essential values that underlie the policy-making process.

Notes

1. The argument in this chapter builds on other work published elsewhere (Gillroy 1992 b).

2. The dominant strategy is that choice that gives a player the best payoffs no matter what choice the other player makes. For example, Row's dominant strategy is R2, because a payoff of 4 is better than 3 (if Column chooses C1) and a payoff of 2 is better than 1 (if Column chooses C2).

3. Much has been written and discussed regarding the matter of whether or not a collective action problem can be solved without central coordination. However, it will be assumed here that the collective action problems involving pollution and especially risk (because of its stealth character) cannot be solved by iteration of the game or by the evolution of convention (Hardin 1982; Taylor 1987). At least there is no evidence that this type of collective action problem could be solved this way and it seems logical that central authority would be needed for monitoring and enforcement, as well as for detection, in most cases.

4. N − k is here assumed to be defectors or free-riders on the collective good. These would impose risk that has not been justified ex ante but has been considered contrary to the public interest. In no way does this contention assume a zero-risk society but one in which all environmental risk imposition is considered (ex ante) on the basis of what it will contribute to individual and/or collective interest and what harm it will bring.

References

Ashby, Eric. 1978. *Reconciling Man with the Environment*. Stanford: Stanford University Press.

Bagley, E. 1961. "Water Rights Law and Public Policies Relating to Groundwater Mining in the Southwestern States." *Journal of Law and Economics* 4: 144–74.

Bish, Charles. 1971. *The Public Economy of Metropolitan Areas*. Chicago: Markham.

Elkin, Stephen L. 1985. "Economic and Political Rationality." *Polity* 18: 253–71.

Elliot, Robert, and Arran Gare, eds. 1983. *Environmental Philosophy*. University Park: Penn State University Press.

Feinberg, Joel. 1973. *Social Philosophy*. Englewood Cliffs, New Jersey: Prentice-Hall.

Feldman, Allan. 1980. *Welfare Economics and Social Choice Theory*. Boston: Martinus Nijhoff.

Fried, Charles. 1977. "Difficulties in the Economic Analysis of Rights." In G. Dworkin, G. Bermantrand and P. Brown (eds.), *Markets and Morals*. New York: John Wiley and Sons. pp. 175–95.

Gillroy, John Martin. 1992 a. "The Ethical Poverty of Cost-Benefit Methods: Autonomy, Efficiency and Public Policy Choice."*Policy Sciences*. 25:83–102.

————. 1992 b. "Public Policy and Environmental Risk: Political Theory, Human Agency and the Imprisoned Rider." *Environmental Ethics* 14:217–237

Goodin, Robert. 1976. *The Politics of Rational Man*. London: John Wiley and Sons.

Hardin, Russell. 1971. "Collective Action as an Agreeable n-Persons' Prisoner's Dilemma." *Behavioral Sciences* 16: 472–81.

————. 1982. *Collective Action*. Baltimore: Johns Hopkins University Press.

Hume, David. 1740. *A Treatise of Human Nature*. L. A. Selby-Biggs (ed.). Oxford: The Clarendon Press.

Lipsey, R. G., and Kelvin Lancaster. 1956. "The General Theory of Second Best." *Review of Economic Studies* 24: 11–32.

Page, Talbot. 1973. *Economics of Involuntary Transfer*. Berlin: Springer-Verlag.

————. 1978. "A Generic View of Toxic Chemicals and Similar Risks." *Ecology Law Quarterly* 7: 207–44.

Pigou, A. C. 1932. *The Economics of Welfare*. Fourth edition. London: Macmillan.

Sagoff, Mark. 1981. "At the Service of Our Lady of Fatima or Why Political Questions Are Not All Economic." *Arizona Law Review* 23: 1283–98.

————. 1983. "The Allocation and Distribution of Resources." Center for Philosophy and Public Policy, University of Maryland, Mimeo.

Snidal, Duncan. 1979. "Public Goods, Property Rights, and Political Organizations." *International Studies Quarterly* 23: 532–66.

Taylor, Michael. 1987. *The Possibility of Cooperation*. Cambridge, England: Cambridge University Press.

Tresch, Richard W. 1981. *Public Finance*. Plano, Texas: Business Publications.

Tribe, L. H. 1972. "Policy Science: Analysis or Ideology?" *Philosophy and Public Affairs* 2: 66–110.

————. 1974. "Ways Not to Think About Plastic Trees: New Foundations for Environmental Law." *Yale Law Review* 83: 1315–48.

2

Moral Domains, Economic Instrumentalism, and the Roots of Environmental Values

Steven Kelman

This chapter will discuss two questions. First, what kinds of values have underlain the growth of environmental concerns over the past decades? Second, is the conflict for policy makers between those environmentalist values on the one hand and, on the other, the approaches toward environmental regulation that come out of economics, as an academic discipline, that in turn dominate much policy analysis about environmental regulation? I will stress the tension between ways that environmentalists look at the world and ways that economists look at the world.

Environmental concerns in some sense date back in American history to before the turn of the twentieth century, particularly to reactions to what from the 1880s into this century people saw as settlement and population moved westward. People in the western parts of the country saw forests being clearcut for timber and left bare, grazing land being overused for sheep or cattle grazing and left naked, disputes in the western states over scarce water resources, and so forth.

One reaction to developments in the western part of the United States was typified by that of one of the founding figures of American environmentalism, Gifford Pinchot. Pinchot was one of the great public managers in American history, the first head of the United States Forest Service, and a person who put his stamp on the culture and mission of the Forest Service in a way that is still alive today (Cooper 1987; Mann 1989; Mitchell 1979, 1990).

Pinchot did not argue that we should not be using forest lands or that we should not be using streams and grazing lands for economic use but rather advocated that we need to be careful and not use these resources so quickly that they will eventually be depleted. The notion of "sustainable yields" in forest management is very much of a Pinchotian notion: One should not cut down forest lands at a greater rate than they regenerate themselves by natural processes. The early conservation movement in the United States had a strong strain of such a view.

At the same time, parallel with the conservation movement, grew a movement that was actually called preservationism," a term still current, associated above all with the name of John Muir (Turner 1985). Muir was born in Scotland, emigrated with his family to the United States in 1849, worked on his family's farm in Wisconsin for a while, and then lost the use of one of his eyes in an industrial accident. After recuperating from the accident, Muir decided to leave factory work and at the age of 29 set out for an exploration, by foot, of over one thousand miles in the southeastern United States. During that trip Muir developed both an appreciation for the non-human creatures he was seeing about him in the wilderness and a contempt for people who did not appreciate the value of other creatures. When he visited the Mammoth Cave in Kentucky, he reacted (according to his diaries) very bitterly against people who lived near the cave but had never visited it because it did not provide anything useful, who thought themselves too wise to waste precious time with weeds, caves, fossils or anything else that he could not eat" (Turner 1985, 134).

The next year Muir headed to California, surprising friends back East who thought that the only reason people went to California was to make quick money. Muir went to explore the wilderness there, and he was bedazzled by his first sight of the Sierra Nevada Mountains in California, and the endless vista that he saw in front of him. He wrote in his diaries that it was here in the valleys of the mountains that he experienced the "floweriest piece of the world I've ever walked, one vast level, even flower bed, a sheet of flowers"(Turner 1985, 164). Muir spent the rest of his life in California, supported himself as a mountain guide, and wrote about the need to preserve unspoiled wilderness.

As early as 1876 Muir wrote that only the protection of the federal government could save the Sierra Nevada Mountains. He wrote in an article that "God over the ages had saved trees from natural disasters but could not save them from fools. Only Uncle Sam can do that"(Turner 1985, 311). In 1889 the publisher of the popular monthly *Century Magazine* recruited Muir as a writer to add his voice for preservation of the wilderness to the other voices for social reform that were featured in the magazine. The publisher wanted Muir to try to set in motion a political campaign to have the Yosemite area of the Sierra Mountains set aside by Congress as a national park. By the time Muir died in 1914 at the age of seventy, he was a man with a well-known face that featured a very craggy, wild beard. And he had become a well-known symbol for an alternative set of values about what made life worth living.

Muir's views broke from traditional American values about unspoiled wilderness. America was born very much in the conquest of wilderness, and there was very little in American tradition that revered unspoiled nature. Muir, after he arrived in California, described the love for the wilderness among his fellow Californians as "almost wholly unknown." Instead, in America untamed wilderness was traditionally seen as wilderness that people had not yet defeated or brought under beneficent human control.

By the end of the nineteenth century, some of that was changing, or beginning to change, in American society. Cities were growing, industry was spreading, many Americans believed that people in the United States were becoming excessively materialistic, and many were upset by what they perceived as the growing power and rapaciousness of big business. This was the period that Mark Twain was to call the Gilded Age. (If he had been writing a hundred years later he might have called it the "Yuppie Age.")

Prior to the Civil War, the federal government claimed much of the western lands, and in the period after the Civil War, much of that land was either given away or sold at very low prices, partly to genuine settlers but often to land speculators and to companies that proceeded to exploit the forests, grazing lands, and streams at a very rapid pace. This exploitation of public lands in the period after the Civil War became a symbol to a significant extent for worries about the general greed and rapaciousness of big business in American culture.

This early strand of American environmentalism was concerned with wilderness preservation, not with issues of health or with other kinds of environmental issues that became important in the environmental movement of the 1970s and 1980s. However, in that John Muir strand of early American environmentalism there were some important strains that one continues to see in the newer environmental movement. One of these is seeing wilderness values as an alternative to urban values and by extension to the kinds of things urban life or industrial life represents. This in turn includes many things. It includes an instrumental approach toward identifying the we care about (nobody values a factory and its smoke in and of itself but only for the kinds of things it produces). And it includes a general view that everything can have a price attached to it. It is therefore not surprising that the early growth of American environmentalism came at a time when people were worried about excessive business influence in society, about too much greed, and about too much rapaciousness.

Let us move forward to the 1960s and 1970s and look at what caused the environmental movement to grow so rapidly (Hays 1959; 1986). In the background was a gradual spread throughout this whole period, but particularly in the postwar era, of the popularity of wilderness experiences and wilderness ideas. It was not just John Muir anymore. For example, in the late 1960s, in the period when environmentalism was getting reestablished, about 46 million Americans held fishing licenses. About 15 million Americans described themselves as people interested in wildlife photography. Samuel Hays (1959) examined guidebooks for mountain hikers. "The earliest ones," he writes, "had stressed mountain climbing. They took the reader rapidly through the terrain, from the starting point to the base of the mountain and then provided detailed directions for climbing them." The important thing was climbing the mountain. As the outdoor recreation movement broadened, the guidebooks began to stress the hiking trail itself, providing details about the various stretches and landmarks to guide the way. Still later, an appreciation of the surrounding biological life came into stronger focus, and guidebooks began to include material on trees and plants in the area through which one was journeying and information about natural history or ecology. With the growth in interest in the wilderness came two new concerns not previously associated with the environmental movement. One was the growing concern over health risks associated with chemicals and toxins in the environment—cancer, asthma, respiratory diseases, and so forth. (These concerns with pollutants people breathe every day actually preceded the later and more current interest in jogging, oat bran, and personal health.)

A second concern that appeared for the first time during the 1960s and 1970s was a more apocalyptic apprehension about ecological balance. At the turn of the century, some people were worried that forests were being clearcut. In the late 1960s and early 1970s, concern spread that because of untrammeled industrialization we were destroying the entire planet—which is much more extreme than just a few forests being cut down. During the late 1960s, there were worries about "limits to growth" and "zero population growth," and, in the 1980s, about the "greenhouse effect."

These issues—the unspoiled wilderness, preserving people's lives and health, and the ecological balance of the planet—at some level might not seem to have a common enough theme to make for a *movement*. One could say that the undermining of these things results from the same cause and that the common theme is that cause. But there is another common theme as well. Valuing wilderness, health, and ecological balance shows a concern about an instrumentalism that makes people willing to do bad things to produce something else that is good and to fail to value sufficiently for its own sake what we are destroying.

The world of the Yosemite wilderness, the Grand Canyon, and other wilderness spaces is, for many people, about as far as one can get from the world of dollars and money making, from the skyscrapers of downtown Manhattan. Similarly, one of the important features of a concern about health and life preservation is the view that human life is not something to which one can attach a dollar sign. Life is something of great value to which no dollar price is attached. And the concern with ecological balance reflects a worry about putting the whole world at risk instrumentally in order to make a buck or to increase production. Thus these different strands of environmentalism all represent a set of alternative values to the values represented by money, by instrumental calculation, by competition, and, in western societies at any rate, by big business.

It is not simply that these environmental values represent an alternative set of values, but also that, to some extent, the values of health protection and wilderness protection are a *repository* for this alternative set of values. When something becomes a repository for something else, part of its value becomes its status as a receptacle for that alternative set of values. It is thus not simply that the Grand Canyon *represents* an alternative set of values to competition, moneymaking, and instrumentalism, but that, in addition, part of the value that people see in the Grand Canyon comes from its *status as a repository* for that alternative set of values.

Except for a tiny fringe within environmentalism, it is hardly the case that most Americans who regard environmental values as important (and there are very large numbers of them, as we know from opinion polls) are hermits, monks, or people who totally renounce all material goods and are totally hostile to commercial values. Connected to the notion of a repository is the idea of *separate domains*. On the one hand, we do not wish to live in hovels, give up television sets or cars, but on the other hand, we at the same time have some worries about those values.

One of the ways to deal with conflicting currents in our minds is to make trade-offs for each decision we make. We decide how much we care about the environment and how much we want a new car, and perhaps we decide to buy a car that gets a little higher gas mileage. Each decision reflects a trade-off among conflicting values. But there is a lot of evidence that people do not always act, or even want to act, that way. In the words of cognitive psychologists James March and Herbert Simon, many people faced with conflicting goals instead pursue a strategy of paying attention to those goals *sequentially* (Cyert and March 1963). In one domain, say the office or the factory, people accept one set of values. But they do not want that set of values to dominate their whole life. So they choose to set aside a separate domain for a different set of values: less competitive, less oriented toward money, less instrumental. The environmental movement has become, I would argue, very much a repository for the desire of people to preserve that separate domain from other things going on in their lives.

To illustrate the plausibility of domain thinking, one may cite an example that has nothing directly to do with the environment, namely, some research by the sociologist Robert Merton (1946). Merton did a study during World War II of a marathon radio war bond drive by singer Kate Smith. Smith did a several-day, non-stop, marathon radio drive to sell war bonds. Merton surveyed people who gave to the war bond drive to determine why they gave and what it was about listening to Kate Smith that motivated them. What impressed many of Mertons respondents was that Kate Smith had made a sacrifice, and they felt that they should make a sacrifice too. Also, Kate Smith made strictly patriotic appeals in her presentation, mentioning nothing about war bonds as a good investment. Respondents were asked specifically whether Smith should have talked about war bonds as a good investment. The vast majority said that she should not and that this would have made them less likely to buy the bonds.

Clearly, much of the advertising in the 1940s, as now, appealed to people's self-interest, including the desire to make good investments. That is certainly not a theme that is foreign to American culture. Why was it, then, that the vast majority of people who gave to the war bond drive did not want Smith to talk about bonds being a good investment? The concept of domain—that people want to keep separate some concerns (not that they do not have the concerns, but they want to keep them separate)—explains what was going on. It was regarded as inappropriate, when talking about helping the country make a sacrifice, to raise money issues. In *that* domain, one should not add a money appeal.

To the extent that this argument is correct—that an important part of the appeal of environmentalism has been its representative or repository status for an alternative set of values—it has created a tension for the formation of environmental policy. This is because a very influential strand of thinking about how we ought to approach environmental regulation in the country comes out of academic economics and policy analysis. The two predominant features of economic analysis as applied to the formation of environmental policy are (1) cost-benefit analysis and (2) what are often called "market-based" approaches toward environmental regulation.

Cost-benefit analysis starts with the observations that we can not spend the whole GNP on the environment and that trade-offs are necessary. Economists go on to say, "Let's spend the amount of money on the environment that will maximize the total net benefits to society." The way to operationalize that is to try to determine, on the one hand, how much it will cost to achieve various levels of environmental controls and, on the other hand, what the benefits of those controls are. To do that an economist will say, "Let's try to find out how much people would be willing to pay to achieve those benefits." The problem that economists encounter is determining the willingness to pay for the Grand Canyon or a scenic vista or saving a life or reduction of the noise level or reduction of odors. They have been very ingenious in dealing with that. Economists have developed all sorts of methods. These methods include taking unsuspecting college undergraduates, giving them a picture of a mountain lake and asking them how much additional a month they would be willing to pay to have that mountain lake outside of their windows? Economists also look at property values in the various parts of Los Angeles and perform complex multiple regression analyses in which they build in controls for everything about the neighborhood except whether it is in a more or less polluted part of Los Angeles. They can then examine the difference in house prices and conclude that differences reflect people's willingness to pay for cleaner air. Having done a calculation of costs and benefits, economists say that society should choose the level of environmental protection at which the benefits of the levels of protection most outweigh the costs.

In market-based approaches to environmental regulation, economists begin with the very correct and important insight that different sources of a given kind of pollution face dramatically different costs in reducing that level of pollution by a given amount (Keese and Schultze 1975; Kelman 1981a). But even within an industry, all things may not be equal. One electric power plant, for example, because it was constructed more recently or for whatever reason, could reduce emissions of sulfur dioxide by a thousand pounds quite inexpensively, while another plant might need to spend three times as much to reduce the emissions by the same amount. Those differences might be even greater across industries: it might be very easy for a certain kind of industry to reduce sulfur emissions, while it is much more expensive for another industry. The economist's argument is that what we really care about is the overall level of emissions, not who reduces them. Therefore, society should put a charge or tax on emissions, or to take a more current effort, society should establish marketable pollution rights (a piece of paper that says the owner of the paper may emit 50 tons of sulfur dioxide), which can be bought and sold in the marketplace. These approaches will not solve the problem, however. If a business can cheaply reduce pollution it will not bother to buy this right because it is cheaper to abate than to buy the right. So the low-cost abaters will reduce their pollution, while the high-cost abaters will buy the right to pollute. The result will be a certain level of emissions reduction, but it will be accomplished by the people who can do it most cheaply.

Both these approaches have at one level a good deal of logic to them. But they also create a tension with environmental values. Take cost-benefit analysis first. One criticism of cost-benefit analysis is the decision procedure that says we should always maximize net benefits to society as a whole rather than making sure we, for example, respect the rights of a small group whose concerns might be trampled on. But I want to put that concern aside here (Kelman 1981b). More important from the point of view of the argument in this chapter is that doing a cost-benefit analysis requires putting *dollar values* on benefits. By requiring that a price be placed on lives, wilderness, and ecological balance, one requires that environmentalists put into dollar terms values that represent separation from the world of dollars. The very process, the very request to those with environmental values that they put those values in dollar terms, is an attack on the values. It precludes health and wilderness from serving as repositories for an alternative set of values to those represented by dollars. Furthermore, making that request to environmentalists reduces their ability to promote the importance of alternative values. Environmentalists wish to go to people and say, "It's not so important to worry about whether I have a BMW or only a Taurus." By saying to environmentalists that they must put all benefits in dollar terms, one undercuts their ability to promulgate an alternative to values that *see* everything in dollar terms. Something similar may be said regarding some aspects of market-based regulation. And this tension between economic analysis and environmental values has, I think, been at the bottom of many conflicts over environmental policymaking and is going to be a challenge for environmental policymakers in the next decade.

We have witnessed over the last twenty years the emergence of a major new cultural and political force in American society. At an institutional level in Washington, we have seen in that time in the politics of organized groups a replacement of the traditional conflict between organized business and organized labor by a conflict between organized business and environmental groups. This is a major change. We saw in the 1980s a very popular president, Ronald Reagan, with a very far-reaching agenda for American society. The one area where he conspicuously failed to move that agenda into public policy was in the area of the environment. The resistance from the public and from elected officials was too great. So in 1988 George Bush promised instead to be the environmental president. On the twentieth anniversary of Earth Day in April, 1990, the cover of *Fortune* magazine declared that the 1990s was going to be "the decade of the environment." In 1993 we have a vice president who has actually written a (best-selling) book about the environment. These issues will continue to be with us and represent a major change in American society over the last two decades.

References

Cooper, John Milton, Jr. 1987. "Gifford Pinchot Creates a Forest Service." In Jameson W. Doig and Erwin C. Hargrove (eds.), *Leadership and Innovation*. Baltimore: The Johns Hopkins Press.

Cyert, Richard, and James March. 1963. *A Behavioral Theory of the Firm*. Englewood Cliffs, New Jersey: Prentice-Hall.

Hays, Samuel P. 1959. *Conservation and the Gospel of Efficiency*. Cambridge, Massachusetts: Harvard University Press.

———. 1986. *Beauty, Health and Permanence: Environmental Politics in the United States 1955–1985*. Cambridge, England: Cambridge University Press.

Keese, Allen V., and Charles L. Schultze. 1975. *Pollution, Prices and Public Policy*. Washington, D.C.: The Brookings Institution.

Kelman, Steven. 1981a. *What Price Incentives? Economists and the Environment*. Boston: Auburn House.

———. 1981b. "Cost-Benefit Analysis: An Ethical Critique." *Regulation* 10: 33–40.

Mann, Dean E. 1989. "Interest Groups and Environmental Policy." In James P. Lester (ed.), *Environmental Politics and Policy*. Durham, North Carolina: Duke University Press.

Merton, Robert K. 1946. *Mass Persuasion: The Social Psychology of a War Bond Drive*. New York: Harper & Row.

Mitchell, Robert Cameron. 1979. "National Environmental Lobbies and the Apparent Illogic of Collective Action." In Clifford S. Russell (ed.), *Collective Decision-Making*. Baltimore: The Johns Hopkins Press.

———. 1990. "From Conservation to Environment Movement." In Michael J. Lacey (ed.), *Government and Environmental Politics*. Washington, D.C.: The Wilson Center Press.

Turner, Frederick. 1985. *Rediscovering America: John Muir in His Time and Ours*. New York: Viking Press.

3

Environmentalism: Values to Politics to Policy

Robert C. Paehlke

Politics, as David Easton taught us in the 1950s, is perhaps best defined as the authoritative allocation of values (Easton 1953). Democratic politics works well, or at least works best, when irreducible value-laden cleavages within society do not dominate the political process. Democracy has trouble with issues like abortion or language rights; they are not easily settled by compromise. As well, democracy frequently cannot establish solid foundations within societies where the overwhelming majority is dramatically less well off than the socioeconomic elite. There must be common ground upon which reasonable compromise can be achieved and widely accepted. This is true of both left-right political divisions and of the newer divisions associated with green politics: environmental protection, economic sustainability, gender equity, and peace.

These latter issues, I will argue, are revolutionizing the pattern of value cleavages in modern societies. But the new conflicts that result need not be irretrievably polarized. Environmental values, through a process of tension and adjustment, can come to be an important and integrated part of our sociopolitical value structure. The conflicts that they engender along the way can and should be resolved. This hope in turn requires that we understand the principle points of tension between environmental values and other sociopolitical value priorities.

Before we look more closely at the value divisions underlying green politics, let me note the simple but highly pertinent argument of the British guild socialist G.D.H. Cole. The question for Cole, writing in the 1930s, was whether it was possible to achieve a thoroughgoing social transformation through a two-party parliamentary system. His

Some material in this chapter derives from research currently funded by the Social Sciences and Humanities Research Council of Canada (SSHRCC).

ambivalence was considerable. A two-party system, he argued, can ensure the "stability and gradual adaptation of an existing system regarded as satisfactory in its essential qualities" (Cole 1933). But it is essential, he continued, that the two rival parties be in close agreement on fundamentals. If they disagreed, other than on lesser matters, they would spend most of their time in office undoing the acts of their predecessor, and society would not get the best out of either alternative system. Even when a parliamentary majority was obtained by a party seeking to transform society, it would be difficult to proceed. Socialist parties "cannot within the brief assured tenure of office which is theirs even attempt by parliamentary means the radical alterations which are necessary for laying the foundations of the new system." Equally, Cole rejected revolution as requiring a widespread desire for revolutionary change and hoped, in Britain, for a third course, one that was neither "purely parliamentary" nor "purely Communist." Cole's argument can be seen to apply equally to the newer value cleavages that are addressed in this chapter.

Cole's case is particularly poignant in the context of the current undoing of the communist states of Eastern Europe and the former Soviet Union. One is again thrown back to the realization that transformational political, social and economic change requires popular support that is both wide and deep. My reading of Cole in the 1960s led me to the conclusion that moderation and tolerance were a sine qua non of political life. Change outside the bounds of democratic politics was unacceptable on these grounds, but democracy was, if real, open to creative adaptation. After the transformations of 1989 and 1990, one sees again in a new light that significant change rarely comes without majoritarian support. One also sees that where such support exists change will often, ultimately, find a way regardless of the details of institutional arrangements. A re-reading of the debates of the 1920s and 1930s is appropriate now because environmental politics involves value divisions that in some ways are as deep as those that motivated theorists like Cole. As well, several contemporary analysts have concluded that environmental politics ultimately poses a genuine threat to democracy itself because a democratic resolution of the cleavages between rich and poor requires prosperity and/or economic growth (Heilbroner 1974; Gurr 1985).

Thus, in the present context there is building within Western societies another social value cleavage of a fundamental sort. Environmentalism as a sociopolitical force is rooted in a profound value transformation. By 1977 Ronald Inglehart had taken some measure of its character and breadth. It has since become broader, and much deeper for some. The environmental historian Samuel Hays (1982) traced the roots of environmental value transformation to the rise of post–World War prosperity and rising educational levels. Environmentalism in his view is thus not simply a response to particular problems "in the environment." It is that, but it is more than that. It is nothing less than a shift in what many human beings consider to be the most important policy priorities. There are increasing signs that an important new dimension in human sociopolitical values is emerging as the twentieth century wanes. In recent publications I have argued that this shift might best be understood as the emergence of a new political ideology—the first since the rise of socialism in the mid-to-waning nineteenth century (Paehlke 1989). These shifts will carry challenges for many political institutions, including a challenge to the very nature and notion of bureaucratic administration (Paehlke and Torgerson 1990).

Environmentalism understood as a political ideology adds a new dimension to the left-right political spectrum. That is, environmentalists frequently reject some basic presuppositions shared by traditional analysts of both the left and the right. Environmentalists, for example, place the highest possible priority on values frequently ignored by political thinkers and leaders of all traditional outlooks—values such as a quality habitat for other species or the long-term availability of ecologically benign energy sources. Also, environmentalists frequently reject goals such as economic growth, indiscriminate technological progress, and national "security" conceived in terms of greater population and military power. But at the same time, environmentalists are variously open to the use of a wide variety of policy instruments traditionally favored by the left (such as taxes, regulation, and even public ownership) as well as those traditionally favored by the right (such as user fees, entrepreneurship, and subsidy removal). This point will be developed further, but the key to understanding environmental politics lies in value analysis and in the complex links between values, politics, and policy. It is to this inquiry that this chapter is directed.

Environmental Values

What, then, are the environmental values that lie at the root of the political shift that is slowly and steadily evolving? *Environmentalism and the Future of Progressive Politics* (Paehlke 1989) sets out a list of thirteen values; there are several other such lists. Mine included an appreciation of the ecological web of life so significant that this value has political salience, a sense of humility regarding the place of the human species within the natural world, a preference for health over wealth when such choices seem to exist, an extended time horizon (concern with the long term in all things, especially as regards the sustainability of the global future), a love of simplicity, a revulsion with waste, a new aesthetic, a non-material basis for prestige and self-esteem, and—surprisingly perhaps—enhanced political democracy.

Inglehart (1977) and others have sought detailed measures of value shifts in these directions under the general heading of postmaterialist values. Dunlap and Van Lierre (1978) and Milbrath (1984) have utilized the general heading of a new environmental paradigm (NEP) in contradistinction to the existing dominant social paradigm (DSP). Milbrath traced the emerging NEP within three nations—the United States, Britain and Germany. One variable—a preference for a society that emphasizes environmental protection over economic growth—was particularly powerful in distinguishing between NEP proponents and DSP defenders (Milbrath 1990, 120). Surprisingly, given the strength of the Green Party in West Germany, the greatest support (5:1) for this view was found in Britain; and while many Germans (2:1) would opt for environment over economic growth, this perspective had more support in the United States. Value shifts do not necessarily find their way directly into party politics or the electoral process. Here, one might suggest, institutions do have a considerable bearing on short-term outcomes (West Germany has a modified proportional representation system).

For analytic purposes, environmental values can be reduced to four core items:

1. The minimization of the negative impacts of human economic activities on ecosystems and habitat
2. The establishment of sustainable patterns of resource use
3. The minimization of negative impacts on human health
4. The maximization of wild spaces and ecological diversity (Paehlke 1990a)

Values at this level of generality are not, of course, adequate guides to policy making. This is because environmental objectives so general will frequently conflict with each other, as well as with non-environmental values. For example, high-yield, yet sustainable forests may lack the species diversity and the flora age range necessary to provide the habitat for many animal species and may offer a setting ideal for species-specific pests. The more we learn about old-growth forests, the more this view seems to be confirmed. Ancient forests provide diversity and resiliency unmatched in managed tracts, however careful the management. Even the protection of human health in a perhaps already overpopulated world can be seen as problematic by some environmentalists. Deep ecology and animal rights in particular have raised some profound disputes within the environmentalist perspective. This self-critical debate has spawned new policy issues and new forms of political action.

But the internal value conflicts of environmentalism are secondary to the problems and the opportunities environmentalism poses in relation to implicit and explicit core values of Western, particularly perhaps North

American, society. The other core values that may conflict with environmental objectives include: (1) technology and science, (2) urbanization, (3) industrialization, (4) economic growth, (5) national security, and (6) social justice. If environmental values are in conflict with all, or even several, of these sociopolitical objectives, one must fear for both environmentalism and the environment. No list could be more politically formidable, especially in a North American context. But environmental values are not straightforwardly in conflict with any of these core social objectives and realities. This conclusion does not imply that there are not real tensions, but does suggest that each of these values will need to be reconsidered in light of the rise of environmental thinking to popular prominence in many political jurisdictions. Let us briefly consider here each of these six values in turn.

Technology, Science, and Environmental Values

Although there is a tension between environmentalism and many modern technologies, environmentalism itself is as much a product of science as an opponent of science and technology. For example, only when the presence of certain chemicals could be detected in parts per million, parts per billion, and parts per trillion did anyone begin to fully understand the effects of these chemicals on ecosystems, food chains, or human health. Environmentalism is a politics rooted in science as no other before it, the claims of Marx and Engels (and Thomas Hobbes and Jeremy Bentham) not withstanding. Ecology, toxicology, epidemiology, climatology, hydrogeology and the several sciences of resource supply (one must be wary here of the term *management*) lie at the heart of environmentalism. Many of the products of technology, too, are greeted with enthusiasm by most environmentalists (e.g., solar photovoltaics, microcomputers, products manufactured from recycled materials, superinsulated houses, energy management systems, and co-generation). Environmentalists seek not the rejection of technology so much as technological selectivity, determined on a basis that integrates explicit democratic value choice and scientific method.

Milbrath (1990) draws a distinction that is particularly apt here. Environmentalists are supportive of science, but not scientism. Science is essential but should not be dominant. Nor should it serve as an implicit prop to established economic and political power.

> We are deluding ourselves if we believe that science is value free. Believing that it is value free has the effect of delivering the control of science into the hands of those holding power and wealth. We need to pay close attention to the values that science serves and redirect its thrust so that it serves society's highest values.

Milbrath argues that modern society accords worshipful deference toward science, where it should instead seek structural ways to critically evaluate what technology produces and how research priorities are selected. Technology, Milbrath asserts, should be subject to ethical controls according to principles recently put forward by Hans Jonas (1984). In summary these involve (1) developing better ways of predicting the long-range effects of technologies; (2) prudently giving greater priority to prophecies of doom over prophecies of bliss; (3) never putting humankind's existence, or basic humanity, at risk; and (4) recognizing our obligations to posterity.

The third of these assertions perhaps needs elaboration here. It might be argued in more general terms that some risks are unacceptable to many environmentalists regardless of the probability of their occurrence. Risk is thus not simply probability times consequence. Some technologies carry very unlikely risks of catastrophic outcomes and may be rejected on grounds that would appear to some analysts to be "irrational." As well, even if the human species itself were not at risk in a moderately more carcinogenic world, some would reject such a risk almost regardless of the costs in economic terms. Others would argue that their very humanity is conditional on the existence of elephants, tigers, and whales living within wild nature. Such value conflicts are deep-seated and will not be resolved without a great deal of forbearance and mutual understanding on both "sides." Arrogant assumptions as to what is or is not scientific are not likely to advance such value disputes. Easy wholesale dismissals of modern science and technology are not helpful either.

Industrialization and Environmental Values

Environmentalists are less opponents of industrialization than products of the postindustrial age. It is no coincidence that environmentalists are most plentiful within the public and service sectors; they are handlers of people, information, and ideas, rather than producers and sellers of products (Cotgrove 1982). But even that generalization may change as more and more environmentally benign (or desirable) products find their way into the marketplaces of the world. Those who earn their living outside the world of manufacturing and sales can more easily come to understand and emphasize the social and environmental costs of ever more things being "consumed" by ever more people. Those within that world often have vague moral doubts but are less often able to embrace the NEP. Milbrath's finding (1990, 55) that trade unionists are three times more likely than business leaders to embrace some key dimensions of the NEP are interesting in this regard. So too are reports that employee morale was greatly improved when a major Canadian supermarket chain chose to emphasize "green" product and packaging initiatives. It is also interesting to note that the United Auto Workers' Black Lake (Michigan) educational center places automobiles at the distant perimeter of the sprawling facility and since the early 1970s has emphasized renewable energy experiments.

One key reason why environmental politics may be coming to the fore in developed economies is that industrial employment has been in rapid decline for decades now. Productivity enhancement over the centuries has reduced the support for traditional left politics. The "gravediggers of capitalism," spawned by industrial capitalism itself, are now a decided and declining minority. But the new majority, while less hostile to capitalism, may be increasingly doubtful about industrialism itself. At the same time, that majority is thoroughly immersed in the quest for the basic products of industry (especially the automobile). Simultaneously, then, individuals recoil from the consequences of collective consumption habits and adhere to many doubtful personal habits on an individual basis. The newest wave of environmentalism (of the late 1980s and early 1990s) would appear to question these habits more directly. Particularly significant, perhaps, are current concerns regarding toxic chemicals, greenhouse gases and forests that may result in considerable changes in transportation habits and in industrial practices and processes. But even these transformations, fully realized, would see changed industries, and perhaps an increasingly postindustrial society, but not necessarily less industry. Less primary industry, less energy and material throughput (Boulding 1973) per dollar of GNP, but not necessarily less industrial employment.

One key to understanding the tensions between environmental values and industrialization lies in two different appreciations of which benefits of industrialization are at risk. Does the reduced use of pesticides threaten basic food supplies or a habitual Western inclination to dandelion-free lawns? Will a contested nuclear power plant light deserted sidewalks and cook dinners or will it power electric toothbrushes and inefficient lighting fixtures? Acceptability of risk is, for environmentalists, principally a function of the prior elimination of all benign alternatives. The relevant environmental merits of particular industrial "advances" cannot be discussed intelligently, in this view, unless a wide variety of efficiency adjustments and obviously helpful changes of habit occur first. Most environmentalists would argue that we can maintain many, if not most, of the benefits of industrial society without necessarily bearing all present costs. Discouraging the trivial uses of energy and materials and improving the efficiency with which industrial society uses energy, materials, land and water is the highest of priorities.

Urbanization and Environmental Values

Equally complex is the relationship between environmental values and urbanization. Many early environmentalists profoundly doubted the desirability of urban conglomerations. Out-of-the-rat-race and back-to-the-land still has some resonance. But there is an urbanist logic to environmentalism as a whole. Cities can be more efficient in terms of energy and materials use per capita than can other forms of human settlement. This could not be more important. More than that, environmentalism itself, as a sociopolitical movement distinct from the conservation movement, has distinguished itself primarily as a movement with urban rather than wilderness concerns: air and water pollution most especially. More recently, the environmental justice movement appears to be carrying this tendency another step further. For example, this movement has emphasized housing concerns as well as the siting (mostly urban) of environmentally undesirable facilities.

While environmentalism still embodies the bucolic dream, many aspects of its logic would, if anything, enhance both urbanization and urban densification. Recycling, for example, works best within urban areas (collection costs are lower). More than that, many aspects of industrial society would be drawn back toward urban centers in the ensuing process changes. Pulp and paper mills that used recycled materials would most often be found in urban, not wilderness, locations. Aluminum smelters are now located near distant hydroelectric sites (to reduce costs), but because recycled aluminum uses only 5 percent of the energy required by virgin materials some smelters might well relocate near the urban sources of materials, supply and product markets. Also, dense cities are particularly energy efficient, both in transportation and space heating terms. Dense cities do not necessarily have less open (green) space—that depends on urban design and especially on the proportion of space used for transportation. Habitat and wilderness are threatened more by sprawl and unregulated wilderness access than by urbanization per se. Being human is perhaps best realized in both urban settings and wilderness settings. In fact, the human preference for urban living, along with some modest limits on public access to the wilderness experience, makes possible the continued existence of the wilderness.

Urban sprawl and rapidly rising human populations carry a significant and ongoing threat to a variety of environmental and non-environmental values. Agricultural lands impinge on wild habitat. Highways replace wetlands, and sprawling urban settlements impinge on agricultural lands. Expanding road systems sever communities and force families and businesses to relocate. All of this results in diminishing wild spaces. The remaining spaces may well come to be more and more highly valued by some. This enhanced, near-to-sacred valuation may obtain for some particular characteristic of the land or because the land is the last wild habitat of some particular species. This change will not, however, necessarily result in rising economic value. Indeed, economic needfulness and willfulness has driven wild spaces back to the very ends of the earth. The only hope for halting this trend is that non-material values will be manifest through the political process as a basis for enhanced protection.

Economic Growth and Environmental Values

There is perhaps in the end (as with industrialization) some measure of irreconcilable conflict between environmental values and economic growth as a societal priority. *Our Common Future* (World Commission on Environment and Development 1987) may be too optimistic regarding prospects of harmonizing environmental protection and economic growth. Some forms of economic growth are compatible with environmental concerns, but the real conflicts are equally important. In a North American (or European or Japanese) context, sustainability and development are as often incompatible as readily reconciled. However, some economic activities would be rapidly expanded in the process of "greening" an economy (energy efficient public transportation and appliances, manufactured goods that use recycled materials, pollution abatement equipment and sewage treatment plants, less chemically dependent agricultural products, etc.). Also, many areas of recent growth are relatively benign per job created. Human services, education, and electronics are, with some exceptions, less problematic than older (smokestack) industries. But many areas of our economy are in conflict with environmental sensitivity: mining and metal smelting, pesticide manufacture and use, some uses of plastics, forestry and pulp and paper manufacture as generally practiced, and, crucially, the automobile. Here is, perhaps, the political agenda of the coming decades: How much is enough? Should we seek to sustain development or to develop sustainability?

The newest impetus to doubts about the excesses of the consumer society has evolved from the inability of North American jurisdictions to site proximate landfills or incinerators for municipal solid wastes. Increasingly we feel, and in reality are, buried in, or surrounded by, our own wastes. Recycling and composting lessen the burden considerably, but rising concern with wastes may also lead to a value shift regarding buying habits. Recent marketing surveys appear to indicate that attitudes about consumption are changing. Another source of this shift, one might speculate, is a key creator of the consumer society: advertising, particularly television advertising. The relentless intrusion of sales efforts may produce, in some proportion of citizens, a deep (or shallow) resistance to consumption itself. It might even be the case that the sales barrage that passes as culture is at the root of rising environmental sentiment.

But what remains questionable is whether changing consumer inclinations necessarily imply a restraint on economic growth. People may purchase more services and fewer goods or fewer more durable and more costly goods. In that case, and others, GNP may rise (and corporate profits may rise), while the physical mass of goods produced declines. Daly (1977) calls this possibility "angelized GNP" and is skeptical about prospects of this sort. However, it is clear that there is a considerable distance to go on many fronts, and enhanced environmental protection does not utterly preclude economic growth. Environmentalists would rarely argue that growth is essential to a good life (in a North American or Western European context), but there are few who assert that GNP growth necessarily precludes either sustainability or environmental protection.

National Security and Environmental Values

Environmentalism is also in conflict with national security, but only as the latter is presently conceived. This conflict exists in part because environmentalists seek a significantly larger allocation of scarce economic resources. But the conflict also exists because the nuclear and chemical weapons industries are themselves among the more significant sources of environmental damage in North America and elsewhere. Conflict also exists implicitly throughout the Third World, where resources are squandered on military imports while ecosystems are destroyed to provide export earnings. The resources devoted to military expenditures in these lands could instead contribute to the mitigation of present ecosystem plights. There is also a conflict at the conceptual/philosophical level. What are the prime components of national security in the present era, knowing what we know of nuclear winter, for example? Does not a greater danger to the security of our nations lie within the realms of ozone depletion, the decline of fossil fuels, the decimation of tropical forests, the greenhouse effect and the spread of toxic air pollution? What is "security"? Increasingly, environmentalists are integrating peace issues within an environmental perspective (see Milbrath 1990, or sources cited in Paehlke 1990a).

Social Justice and Environmental Values

Finally, what of conflicts between environmentalism and social justice? This is a topic that requires books, many of them, not just a few paragraphs. But here the conflicts are less irreconcilable, though nevertheless real. There are, for example, more, not fewer, employment and economic opportunities within a green economy, but they are not the same opportunities or necessarily opportunities in the same places at the same wage levels. For example, there are more jobs in recycling, reforestation and energy efficiency than in equivalent tree harvesting or energy megaprojects (Paehlke 1989). In addition to employment issues, it might be observed that pollution, locally unwanted land uses, and occupationally based health hazards are not randomly distributed. Less advantaged economic groups face a disproportionate share of environmental problems (see sources cited in Paehlke 1989, chapter 2, Bullard 1991, and Blumberg and Gottlieb 1989). On the other hand, there is a potential political conflict between social justice advocates and environmentalists over relative shares of governmental program funding.

This latter tension would, however, seem to be as much an opportunity for, as a limitation on, environmental advance in the political realm. For example, the poor spend a disproportionate share of their income on energy. Public funding of home insulation or efficient appliances for the elderly poor would be both socially and environmentally beneficial. Also, enhanced educational and employment opportunities for women correlate with decisions to restrain family size. Similarly, the employment opportunities inherent in some environmentally beneficial initiatives (e.g., public transportation subsidies or renewable energy initiatives) could be targeted to a high proportion of gender-specific or minority employment. Given that environmentalists are relatively unenthusiastic about unchecked, unmanaged, undirected economic growth, it would be politically astute if they were sensitive to the needs of less advantaged groups. Ignoring such questions has in the past left environmentalists vulnerable to charges of elitism (as in Tucker 1977).

Politics as Value Compromise

How, then, does society move beyond the development of environmental values and achieve an integration of these new values with the old? Indeed, how is a better relationship between humanity and the environment actually achieved? In this there is no avoiding politics, the slow steady conflicts and compromises, gains and losses, ups and downs, of democratic change. In particular, there is as well no avoiding the complexities of between-election bureaucratic decision making. One cannot any longer, if one ever could, just escape the damage and grow organic vegetables. One soon finds environmental troubles blowing in one's wind or sited upstream from one's water supply. Increasingly, in any case, environmental damage is global damage. One need only mention here acid precipitation, global warming, damage to the ozone layer and generalization of toxic contamination.

The politics of the environment urges a rethinking of the ethos and practice of capitalism, socialism and the variety of mixed economies alike. From Thomas Malthus, William Jeavons, and G. P. Marsh to John Muir, Henry David Thoreau, and Aldo Leopold, the intellectual building blocks of environmentalism were put in place and public concern evolved. Since the 1950s environmental politics has ebbed and surged with each peak seemingly more intense than the previous one. Presidents and prime ministers must now claim not only that they have come to an appreciation of these concerns but at times do even claim that they have always been environmentalists. Membership in mainstream environmental organizations, such as the Sierra Club, has grown steadily through the 1980s. They, and other such organizations, employ more and better lobbyists, attorneys and organizers. Newer, more environmentally radical organizations have grown even faster. Greenpeace has nearly three million contributors in North America. Green consumer products may be the most important marketing innovation to hit Britain and Canada since the advent of television. Green parties may yet alter the traditional, established electoral patterns of both Western and Eastern Europe in this decade.

Beyond partisan politics, environmentalism is having a significant impact on the theory and practice of public administration. Administrative practice is opening up to advocates other than business, labor and regional interests. In dozens of ways citizens are involving themselves in the daily decision-making processes of bureaucratic government at the state and municipal, as well as the federal, level. Increasingly, environmental considerations are also finding their way directly into economic decision making through non-governmental or quasi-governmental paths. Increased costs for energy supply and waste disposal have this effect, sometimes enhanced by policy action. So, too, do environmental audits (some of which go well beyond matters of regulatory compliance). So, too, does direct bargaining with industry at the community level or the integration of an environmental dimension into labor negotiations. The active effort on the part of individual and institutional consumers to buy environmentally friendly products can have important effects, as can the characterization of investment opportunities as more or less desirable in environmental terms. All of these initiatives are presently taking place in North American jurisdictions. The participatory dimensions of environmental politics are highly diverse.

Finally, what policy imperatives as regards the management and regulation of environmental risk are implicit in the foregoing political developments? One implication is that regulation itself is coming to be seen by some as a necessary but not sufficient means to achieve environmental protection. Another way to characterize this new approach is as an all-tools, all-levels, all-sectors approach to environmental protection (Paehlke 1990b). That is, environmental protection can be achieved using governmental procurement policies, taxation (on carbon or oil or energy), user fees (on waste pickup or time-of-day freeway use), subsidy or tax expenditures (on investments in materials reuse or energy efficiency improvements), subsidy removal (on primary goods production), moral suasion, or good old-fashioned sticks (criminal law). The full range of tools available to government can be used creatively in the achievement of environmental objectives. An all-sectors approach might produce the application of environmental audits, for example, to universities or government agencies, or the application of environmental impact assessment to the running of a federal deficit. An all-levels approach could result in the creative use of municipal government, especially regarding waste reduction and the control of carbon emissions (fuel use). For example, the city of Toronto recently considered prohibiting the sale on city property (sidewalks) of newspapers printed on non-recycled paper. That city is also now considering restrictions on automobile entry to the downtown core as a means of lowering carbon dioxide emissions by 20 percent.

Environmental Values and Risk Management

This altered approach to environmental policy is not yet firmly established. But if it is a coming trend, there are several important implications for the overall management of environmental risk. First, risk management will not be easily confined to one or two particular agencies, such as the Environmental Protection Agency. Second, risk management may be indirect in character as often as direct. A general government services agency may procure recycled paper in order to strengthen the market for curbside recycling collection. More recycling in turn will reduce industrial energy demand and thereby reduce the risks attendant on energy production, be it from coal, nuclear, or other energy sources. In brief, the new approach is non-regulatory and accepts the proposition that all energy and material thoughputs are potential environmental problems. The overall policy objective of increasing GNP output per unit of throughput supplements in an important way the regulation of particular risks. Third, economic markets are empowered as risk management devices through the internalization of environmental externalities. However, it is arguable that markets ought not be a substitute for authoritative decision making in cases where there are serious immediate risks to human health or environmental quality. Carcinogenic chemicals and endangered wetlands are risks that can only be managed directly through the power of law and administration. To recognize the usefulness of some economic devices in some circumstances is not to necessarily deny the need for political means in others. The process whereby environmental values proceed through the political to the policy realm will determine which policy devices are appropriate in which circumstances. Risk assessment and environmental impact assessment are two means of aiding the process of value integration.

Once it is granted that some risks must be handled by political, legal and/or administrative means, there are two additional considerations that can be offered in conclusion. First, the assessment of risk must be undertaken within longer time frames, with a more frequent and more empathetic consideration of non-human species. In brief, the non-acceptability of risk must become more extensive in time, space and species terms. Why, one would ask from this new perspective, risk the beluga whales of the Beaufort Sea to achieve but a two-year supply of oil for North America? If our society and economy must do without conventional fossil fuels shortly in any case, how important is so brief an increment of time? This perspective becomes apparent if one's analysis is oriented in decades or centuries, rather than months or years.

Second, to reiterate and restate another of Jonas's (1984) points, integration and assessment must emphasize the dangers in irretrievable steps. We cannot ever restore old-growth forests or prairie grasslands, especially if none remain. Toxic chemicals, introduced into the environment, are almost impossible to recover. There is likely no completely satisfactory alternative to conventional oil and natural gas, and there may never be. On the other hand, we can bring wolves back to Yellowstone Park and bears to some European forests. We might even be willing and able to restore a few lions in Great Britain. But we cannot easily replace a tropical forest or a coral reef. Some steps are irretrievable. Many irretrievable steps have already been taken, or are about to be taken, some on a planetary scale. Only if environmental values prevail in the political process can human security and comfort exist without such destructive acts against nature.

References

Blumberg, Louis, and Robert Gottlieb. 1989. *War on Waste.* Washington, D.C.: Island Press.

Boulding, Kenneth E. 1973. "The Economics of the Coming Spaceship Earth" in Herman E. Daly, ed., *Toward a Steady-State Economy.* San Francisco: W.H. Freeman

Bullard, Robert D. 1991. *Dumping in Dixie: Race, Class, and Environmental Quality.* Boulder: Westview Press

Cole, G.D.H. 1933. *The Intelligent Man's Guide Through World Chaos.* London: Gollancz.

Cotgrove, Stephen F. 1982. *Catastrophe or Cornucopia: The Environment, Politics and the Future.* New York: Wiley.

Daly, Herman E. 1977. *Steady-State Economics.* San Francisco: W. H. Freeman.
Dunlap, Riley, and Kent Van Lierre. 1978. "The New Environmental Paradigm." *The Journal of Environmental Education* 9:10–19.
Easton, David. 1953. *The Political System.* New York: Knopf.
Gurr, Ted. 1985. "On the Political Consequences of Scarcity and Economic Decline." *International Studies Quarterly* 29:51–75.
Hays, Samuel P. 1982. "From Conservation to Environment: Environmental Politics in the United States Since World War Two." *Environmental Review* 6:14–41.
Heilbroner, Robert. 1974. *An Inquiry into the Human Prospect.* New York: Norton.
Inglehart, Ronald. 1977. *The Silent Revolution.* Princeton: Princeton University Press.
Jonas, Hans. 1984. *The Imperatives of Responsibility: In Search of Ethics for the Technological Age.* Chicago: University of Chicago Press.
Milbrath, Lester. 1984. *Environmentalists: Vanguard for a New Society.* Albany: SUNY Press.
————. 1990. *Envisioning a Sustainable Society.* Albany: SUNY Press.
Paehlke, Robert. 1989. *Environmentalism and the Future of Progressive Politics.* New Haven: Yale University Press.
————. 1990a. "Environmental Values and Democracy: The Challenge of the Next Century." In Norman J. Vig and Michael E. Kraft, eds., *Environmental Policy in the 1990s.* Washington, D.C.: Congressional Quarterly Press.
————. 1990b. "Toward a New Environmental Protection Strategy: A Survey of Options with an Emphasis on Government Procurement." *Business in the Contemporary World* 2:30–37.
Paehlke, Robert, and Douglas Torgerson. 1990. *Managing Leviathan: Environmental Politics and the Administrative State.* Peterborough, Ontario: Broadview Press.
Tucker, William. 1977. "Environmentalism and the Leisure Class." *Harper's* 255:49–56, 73–80.
World Commission on Environment and Development. 1987. *Our Common Future.* New York: Oxford University Press.

PART TWO

Value Conflicts, Domain Trade-offs, and Political Cooperation

4

The Nature of Environmental Values

Vicente Medina

In this chapter I assume that there is a moral conflict between the maximization of profit and pleasure, on the one hand, and the preservation and promotion of a healthy environment, on the other.[1] In light of this assumption, I would like to explore the extent to which *environmental objects*[2] are valuable and whether all, or at least some, of them have intrinsic (noninstrumental) as well as extrinsic (instrumental) value. In order to illuminate the distinction between intrinsic and extrinsic value I will refer to Arne Naess's distinction between "shallow" and "deep" ecology (Naess 1973).

I hope my argument will clarify several issues. First, the distinction between intrinsic and instrumental value is independent of that between shallow and deep ecology. Second, the distinction between shallow and deep ecology is not necessarily related to the controversy about the anthropocentricity of values. And third, the distinction between intrinsic and instrumental value is not directly related to that between the objectivity and subjectivity of values (Moore 1922).

If we discover that all or some environmental objects are intrinsically as well as instrumentally valuable, then to the extent to which moral principles and values govern our interaction with other moral agents, they should also govern our interaction with the environment. If an environmental object X is valuable (intrinsically or instrumentally good), then any moral agent Y that is capable of affecting X has a prima facie obligation[3] to act compatibly with the preservation of X. But if there is ever a serious moral conflict between the welfare of actual or prospective moral agents and the welfare of environmental objects, the welfare of the former outweighs that of the latter. Moral agents come first not because our ethics are necessarily anthropocentric but because moral agents are more valuable than environmental objects by virtue of their nature.

I would like to thank David O'Connor, David Benfield, and Ramon Lemos for their constructive criticism.

Environmental Dilemmas and Human Ethics

It seems evident that with our new technological developments we are undermining and hence risking the life of present and future generations. For example, consider the problems of the depletion of the ozone layer, the depletion of natural resources, the disposing of toxic waste such as nuclear waste, the contamination of our air and drinking water, the contamination and destruction of our shores by oil spills and the dumping of toxic waste materials, and the destruction of ecosystems in general by our lack of knowledge or by irresponsible behavior on the part of some individuals. Thus it is reasonable to assume, as John Passmore contends, that if we want to maintain a sound quality of life for ourselves and for future generations we need to change our behavior (1974, 3).

If we want the human species to survive, we do not have any choice but to develop a socially responsible attitude or perhaps a new "ecological attitude" (Blackstone 1974), thereby changing our ways of interacting with the environment. The issue is precisely the price we are willing to pay in order to bring about these changes. Here is where Passmore disagrees with some ecologists, like Aldo Leopold, who argues that what the West needs is a "new ethic" (1974, 3–4).[4] Whether this is the case remains to be seen. Perhaps what we need, as Frankena (1979, 3-4) ably argues, is not a "new" ethics but instead a commitment to morally responsible behavior in our relationship with the environment.[5]

Is our Western economic tradition of maximizing pleasure for the present generation morally consistent with our concern for a healthy environment not only for present but for future generations? The answer is not so obvious as it appears. It will depend on how one interprets these two contestable terms: *pleasure* and *healthy environment*.

If one interprets the term *pleasure* as to mean the satisfaction of desires without taking into consideration the possible detrimental effects that our actions may (directly or indirectly) have on other people's lives, then the answer to the question is clear. The maximization of pleasure at the expense of other's well-being is unfair and hence morally impermissible because this behavior violates a fundamental moral principle, namely, *we ought not to harm others without sufficient justification*. I do not think the promotion of pleasure a *sufficient justification* for seriously undermining or risking the lives of others. However, if one understands the term *pleasure* as referring to satisfaction of only "positive desires,"[6] then the answer is not so clear.

Similarly, if one characterizes the concept of a healthy environment as one that provides the necessary conditions for satisfying the desires of present generations while seriously undermining or risking the lives of future generations, the answer is clear. This is simply unfair and hence wrong for two main reasons: (1) we may be directly harming some members of the present generation, and we are likely (directly or indirectly) to harm future generations, and (2) as members of the present generation, we are taking advantage of our privileged position by taking actions that benefit ourselves but, most likely, will harm future generations without their consent.[7]

We may be able to circumvent the above objections by characterizing the concept of a healthy environment as follows: a particular environment E is healthy if and only if there are sufficient environmental objects in E for the preservation and promotion of some kind of life. I am leaving out the question of what "kind" of life ought to be preserved and promoted because I will discuss that later.

Nonetheless, regardless of how one defines the concept of a healthy environment, it can never be a risk-free environment. Even a radical ecologist like Arne Naess acknowledges that "any realistic praxis necessitates some killing, exploitation, and suppression" (1973, 95). The same concept applies to risks taken in society. As Passmore argues, what we should demand is not a risk-free society, because there is no such thing, but an "account of the ecological costs and benefits of [our] actions" (1974, 50). Moreover, I would argue for responsible moral behavior in our interaction with the environment. This means that we have a *serious moral obligation*[8] to act compatibly with the preservation and promotion of a healthy environment not only for our own good but for the good of "others"[9] who are affected by our actions.

A New Environmental Ethics?

Do we really need, as some ecologists contend, a new "environmental ethics," "environmental philosophy," or "ecosophy" (Naess 1973, 99) that will take the place of traditional Western ethical theories? In order to answer this question, I will explore the accuracy of and the degree of illumination provided by the distinction between shallow and deep ecology.

According to some ecologists, there is an important distinction between shallow and deep ecology (Naess 1973, 95).[10] Shallow ecology is characterized as being essentially anthropocentric. That is, shallow ecologists normally argue that environmental objects are valuable to the extent that they contribute to human well-being. Some deep ecologists take this to mean (1) that humans are "the source of all value" and (2) that the environment has only instrumental value in relation to humans (Fox 1984a, 194).

This way of characterizing the shallow view is misleading. First, at least some traditional Western moral theories do not necessarily refer to humans as the source of all value but rather to "moral agents," and whether there are nonhuman moral agents remains an open question.[11] Second, it is also misleading to suggest that Western moral theories converge on the nature of value. Sometimes, depending on the theory, values are described as relational properties (instrumental) or as monadic properties (intrinsic) or as both. In some theories, whether there are intrinsic values does not necessarily depend on the existence of humans or any form of consciousness but may instead depend on the existence of certain properties that may or may not be related to the existence of some form of consciousness.[12]

Similarly, the question of whether values are only instrumental is not necessarily related to the existence of humans but instead to the existence of moral agents or some form of awareness to which these values contribute. There is nothing necessarily anthropocentric about this last claim. To argue, as some environmentalists do, that this position amounts to "class chauvinism" or "human chauvinism" is simply to miss the point and the complexity of the issue in question.[13]

Deep ecology is depicted as non-anthropocentric in recognizing the intrinsic value of environmental objects and their right to exist in order to fulfill their own "evolutionary destinies" (Fox 1984a, 194; Naess 1973, 96). This position does not exclude Western ethicists from being committed to it. This would depend on how one would interpret the concepts of "intrinsic value" and the "right to exist." The issue about the meaning and existence of intrinsic value is separate from the one about the anthropocentricity of value. One does not confer or discover values simply because one is a member of the class of Homo sapiens but because one has the capacity for reasonable evaluation.

In order to elucidate the controversy about the existence of intrinsic value, I shall address the distinction between the objectivity and subjectivity of value. But first, what is a "value?" I am not asking about which things or objects are valuable but rather about what the term *value* refers to when it is understood in abstraction from any specific valuable object. I characterize the term *value* in this way: V is a value if and only if V denotes some positive attribute, property, or quality of particular states of affairs, objects or actions so that if a "rational moral agent"[14] were in a position to choose between F and G, and V is present in F but absent in G, she would choose F rather than G based on the actual presence of V in F.

There are several ways one may characterize the distinction between objective and subjective value. One way of doing so is to refer to an epistemological distinction between "agent-relative" and "agent-neutral" reasons. An agent-relative reason is, roughly speaking, a justification a moral agent presents for doing A or desiring B on the ground that A or B is compatible with his or her interest(s). On the contrary, an agent-neutral reason is, roughly speaking, a justification a moral agent would present for doing A or B regardless of whether A or B would promote his or her interest(s) (Benn 1988; Nagel 1980; Wiggins 1976). If one accepts the latter, then one is committed to some form of the "moral point of view" or to the principle of "universalizability."[15]

One may also interpret the dispute about the objectivity or subjectivity of value as a dispute between those who argue that there are objective value-properties independent of our acts of valuing or preferring that may, in principle, be experienced by any actual or prospective moral agent, and those who argue, on the contrary, that values are simply the subjective preferences of an individual or a collection of individual moral agents. Yet, as G. E. Moore (1922) contends, the distinction between objective and subjective value is different from the distinction between *intrinsic* and *instrumental* value.

One may accept the view that there are objective moral values and disvalues and deny without inconsistency that there are any intrinsic values or disvalues.[16] But once we acknowledge the existence of intrinsic value as that which is valuable for its own sake, we are committed to the objectivity of values and disvalues. Thus to say that something Q has intrinsic value or disvalue is to say that Q's value or disvalue depends solely on Q's nature rather than on Q's causal efficacy or relationship to something else.[17] To say that something R has extrinsic or instrumental value is to say that R's value does not depend solely on R's nature but instead on R's causal efficacy or relationship to something else that may also be instrumentally or intrinsically valuable. Accordingly, to say that something S has extrinsic or instrumental disvalue is to say that S's disvalue does not depend solely on S's nature but instead on S's causal efficacy or relationship to something else that may also be instrumentally or intrinsically disvaluable (not valuable).

If we agree with deep ecologists that environmental objects in general, including the universe itself, are valuable independent of any moral agent or any form of awareness of them, are we also necessarily committed to maintaining that we have a serious obligation to protect these environmental objects at the expense of considerably undermining or risking the life of some actual or prospective moral agents? I do not think so. One may accept the objectivity of value without necessarily embracing "value egalitarianism" (the view that *all* values have equal worth).

Unlike environmental objects, moral agents are valuable for at least three main reasons: (1) they have the capacity to experience intrinsically valuable states of consciousness, for example, happiness, pleasure and knowledge (to mention only three possible candidates), (2) they have the capacity to recognize values and disvalues in the universe independently of their appreciation of them, and, more important, (3) they are evaluators, that is, they have the capacity not only to recognize but to confer value as well.

If what I have just said is correct, then the question of whether humans are the source of all value is simply moot. The important issue is not whether one is willing to embrace the principle of "biospherical egalitarianism"—that every form of life has "the equal right to live and blossom" (Naess 1973, 95–96)—but whether one is willing to accept the view that *all* forms of life are equally valuable.[18] Apparently, this is not what some deep ecologists have in mind. Even Naess suggests that, on practical grounds, there must be some sort of trade-off (1973, 95; Fox 1984a, 199). This, it seems to me, implies that even though deep ecologists recognize all forms of life as having an equal prima facie right to exist, they also recognize that some of them are more valuable than others. Therefore, if they do not want to be incoherent, they must accept some kind of perfectionism.[19]

One may reasonably maintain that every environmental object has a prima facie right to exist. That is, in principle, we have a prima facie negative obligation not to destroy any environmental object without sufficient justification (Lemos 1986, Chapter 3). If this is so, then I argue that environmental objects have a prima facie right to exist only if and so long as (1) their existence does not considerably undermine or risk the life of either more valuable environmental objects or that of moral agents and (2) their existence does not prevent that of more valuable environmental objects or that of moral agents.[20] Moreover, we have a prima facie positive obligation to protect environmental objects only if and so long as (1) they have intrinsic value or (2) they contribute to the well-being and self-realization of moral agents.

This means that when there is a serious moral conflict between the welfare of actual or prospective moral agents and the existence of environmental objects, the first outweighs the latter, not because our ethics is necessarily anthropocentric, but because moral agents are more valuable than environmental objects by virtue of their nature.[21] Even if we agree that some environmental objects have intrinsic value, this does not commit us to the view that environmental objects and moral agents are equally valuable and hence have an equal *actual right to exist*. We are simply committed to the view that both have an equal prima facie right to exist. However, they do not all have an equal *actual* right to exist. Even in the case of environmental objects, some are more valuable than others by virtue of either their intrinsic value or their instrumental value in promoting the welfare of moral agents. Whenever there is collision of rights and obligations our duty is to fulfill our weightiest obligation in that particular situation.[22]

Conclusion

If one is willing to embrace the view that some environmental objects have intrinsic as well as instrumental value, then one may also accept the view that we have both a prima facie negative obligation not to destroy these objects without sufficient justification and a prima facie positive obligation to preserve them. As philosophers, we may disagree on the nature and ontological status of values. Yet, as responsible citizens of a democratic state, we should acknowledge that even though the question about the nature of these objects is open-ended there seems to be a collision among them in society. Liberty, justice, equality, beauty, happiness, and health (to mention only a few) are all competing with one another. Hence, it is difficult to say which one takes precedence over the rest in all possible situations.[23]

Regardless of how one interprets the notion of value, we have a serious moral obligation to teach respect for environmental objects to the extent they are necessary for the preservation of a healthy environment and hence for the promotion of our health and that of future generations. To try to reduce environmental problems to cost-benefit analyses (one version of the shallow view) or to the obscure notion of biospherical egalitarianism is simply to miss the complexity of environmental problems by presenting a false dilemma. The shallow and the deep views share at least some common ground, namely, the recognition of one's moral obligation to preserve and promote a healthy environment.

Notes

1. This may be seen as a loaded assumption or as a moral condemnation of capitalism. But it seems to me that the moral tension that exists between the maximization of profit and the preservation and promotion of a healthy environment is similar to the one that exists between satisfying desires and fulfilling our moral obligations. If this is the case, then trade-offs are unavoidable in society.

2. I am using the term *environmental objects* to refer to all kinds of natural objects (including rivers, trees, landscape, wilderness, mountains), all kinds of living organisms, and microorganisms, and ecosystems in general.

3. I am using the term *prima facie obligation* in the same way W. D. Ross (1930) used the term *prima facie duty* or *conditional duty*. The following definition captures what I mean by a prima facie obligation. A moral agent Y has a prima facie obligation to do or bring about Z if and only if Y ought to do or bring about Z, unless doing so would conflict with fulfilling some weightier obligation.

4. Leopold (1987) advocates extending the realm of ethical values to encompass our relationship with the "biotic community" and the recognition of the intrinsic value of this community. For similar views, see also Rolston (1974; 1987); Taylor (1986); Hargrove (1989); Callicott (1986; 1990).

5. It seems to me that even if we do not need a "new" ethics, a good case could be made for some kind of reevaluation of our relationship with the environment. To this extent, we should be thankful to environmentalists for presenting their cases not only as philosophical problems but, more importantly, as social policy issues.

6. I am using the term *positive desires* to refer to those desires that are not in conflict with the well-being of others and consequently do not violate important moral rights.

7. This may seem trivial, because members of future generations are just prospective rather than actual moral agents. Therefore, to raise the issue of consent in this context may seem to be otiose. Yet I do not think this is the case, because one way of taking seriously the well-being of future generations is to ask the question, Would they have approved or consented to our present actions if they had known about the possible detrimental effects upon them? I understand the objection that hypothetical people are not actual people and hence hypothetical rights are not actual rights. Nonetheless, by putting ourselves in other people's shoes we are adopting the so-called moral point of view, and this seems to be a reasonable choice (Rawls 1971, 118–92).

8. A "serious moral obligation" is, according to my account, weightier than a prima facie obligation because it emerges as a result of our interaction with the environment or with other moral agents in a life-and-death situation.

9. In this context, the concept of "others" refers primarily, although not necessarily, to moral agents. But it may also be used to refer to natural objects if their existence is not in serious conflict with the existence of moral agents.

10. See also Fox (1984a), Naess's sympathetic reply to Fox's article (Naess 1984) and Fox's rejoinder (1984b).

11. I am using the term *moral agents* to refer to "persons" in the moral sense rather than in the genetic sense. A being X is a person if and only if X possesses some characteristics or properties by virtue of which he or she is a full-fledged member of the moral community. The emphasis on "moral agency" is on the capacity to make reasonable decisions for one's life and, to that extent, to distinguish right from wrong behavior. From a metaphysical point of view, there is nothing in the connotation of the term "moral agency" that would restrict its application only to humans. If this term were to be restricted only to humans, then I would agree with deep ecologists that to reduce values only to anthropocentric considerations would be simply arbitrary and arrogant on the part of traditional Western ethicists. For an illuminating discussion on personhood, see Feinberg (1986).

12. For a respectable tradition embracing the "objectivity" of values as having independent ontological status, see Moore (1903); Ross (1930); Brentano (1969); Meinong (1972); Scheler (1973). And a recent and original defense of this view, see Butchvarov (1989).

13. For a critique of some Western ethical theories of values, see R. and V. Routley (1979).

14. I am assuming that "rational moral agents" are those individuals who have the capacity to distinguish and thereby choose between a right and wrong,

good and bad, valuable and disvaluable (not valuable) state of affairs. Moreover, it seems to me that the concept of "rationality" involves more than just the capacity to maximize wants or desires. It also involves assessing the moral qualities of these wants or desires. For some further comments on the concept of rationality, see my book (Medina 1990, 106–07).

15. When one adopts this point of view, one is at least committed to applying important moral principles or rules consistently regardless of who is affected by them. See Baier 1965; Frankena 1973; Kant 1959; Hare 1981; Rawls 1971.

16. There is no necessary inconsistency here because one may interpret "objective values" as referring to "agent-neutral" reasons for action instead of referring to "intrinsic values." See Nagel (1980, 77–139).

17. This position is similar to Moore's (1922, 260). For a similar distinction between positive and negative value or what I call value and disvalue, see Scheler (1973, 26).

18. If one rejects perfectionism, that is, the view that some forms of life are more valuable and hence preferable to others, then one must be ready to back this up with some reasonable arguments rather than, as in the case of some deep ecologists, with an appeal to intuition.

19. This, as Andrew Dobson contends, is a real problem for deep ecologists, because from their point of view one does not have any clear criteria of how to go about preserving and promoting some species and killing and exploiting others. The interesting question is how to accomplish this trade-off among environmental objects. See Dobson (1989).

20. It is true that my view may commit me to the farfetched example that if one were in a lifeboat situation and had to choose between the life of a morally despicable person, one who had committed crimes against humanity, and that of a kind animal, one would be obliged to save the person instead of the animal. But a possible reply to this example is that one would be obliged to save the despicable person not because he is more valuable than the animal but because one has an obligation to bring him to justice. Moreover, I am willing to maintain that the value of justice overrides the value of our poor animal. See R. and V. Routley, (1979, 57).

21. On this point I agree with both J. Passmore (1974) and H. J. McCloskey (1983, 29–61). However, my position differs from theirs because I am using both a concept of moral agency and a concept of value that are not necessarily anthropocentric.

22. I am using the term *duty* as E. F. Carritt (1973, 3) used it, to mean "the strongest present obligation" in a given situation. The term *duty* may also be understood in a consequentialist way as that act by which one brings about a greater balance of value over disvalue in a given situation.

23. For an illuminating defense of value pluralism, see Isaiah Berlin (1969). It seems to me that besides the question of whether environmental ethics must be committed to "moral pluralism" or "moral monism" in the metaphysical sense, which I think is an open question, we are still faced with the problem of value pluralism and hence the inevitability of trade-offs among values. This is precisely the challenge for any ethical theory. For a recent version of moral monism in environmental ethics, see Callicott (1990).

References

Baier, Kurt. 1965. *The Moral Point of View*. New York: Random House.

Benn, Stanley I. 1988. *A Theory of Freedom*. Cambridge: Cambridge University Press.

Berlin, Isaiah. 1969. "Two Concepts of Liberty" in *Four Essays On Liberty*. Oxford: Oxford University Press. pp. 118–72.

Blackstone, William T. 1974. "Ethics and Ecology" in *Philosophy and Environmental Crisis* ed. William T. Blackstone. Athens, Georgia: University of Georgia Press. pp. 16–42.

Brentano, Franz. 1969. *The Origin of Our Knowledge of Right and Wrong*. New York: Routledge & Kegan Paul.

Butchvarov, Panayot. 1989. *Skepticism in Ethics*. Indianapolis: Indiana University Press.

Callicott, J. Baird. 1986. "The Search for an Environmental Ethic in Matters of Life and Death" in *New Introductory Essays in Moral Philosophy*. ed. Tom Regan, 2nd ed. New York: Random House.

———. 1990. "The Case Against Moral Pluralism." *Environmental Ethics* 12: 99–124.

Carritt, E. F. 1973. *Ethical and Political Thinking*. Westport: Greenwood Press. [Reprint of 1947 Oxford edition.]

Dobson, Andrew. 1989. "Deep Ecology." *COGITO* 3: 44–45.

Feinberg, Joel. 1986. "Abortion" in *Matters of Life and Death*, ed. Tom Regan. New York: Random House. pp. 258–70.

Fox, Warwick. 1984a. "Deep Ecology: A New Philosophy of Our Time?" *The Ecologist* 14: 194–200.

———. 1984b. "On Guiding Stars to Deep Ecology." *The Ecologist* 14: 203–04.

Frankena, W. K. 1973. *Ethics*. Englewood Cliffs, New Jersey: Prentice-Hall.

———. 1979. "Ethics and the Environment" in *Ethics and Problems of the 21st Century*. ed. K. E. Goodpaster and K. M. Sayre. Notre Dame, Indiana: University of Notre Dame Press.

Goodpaster, K. E., and Sayre, K. M. eds. 1979. *Ethics and Problems of the 21st Century*. Notre Dame, Indiana: University of Notre Dame Press.

Hare, R. M. 1981. *Moral Thinking*. Cambridge: Cambridge University Press.

Hargrove, Eugene C. 1989. *Foundations of Environmental Ethics*. Englewood Cliffs, New Jersey: Prentice-Hall.

Kant, I. 1959. *Foundations of the Metaphysics of Morals*. Indianapolis: The Bobbs-Merrill Co.

Lemos, Remon M. 1986. *Rights, Goods and Democracy*. Newark, Delaware: University of Delaware Press.

Leopold, Aldo. 1987. *A Sand County Almanac and Sketches Here and There*. New York: Oxford University Press.

McCloskey, H. J. 1983. *Ecological Ethics and Politics*. Totowa, New Jersey: Rowman and Littlefield.

Medina, Vicente. 1990. *Social Contract Theories: Political Obligation or Anarchy?* Totowa, New Jersey: Rowman and Littlefield.

Meinong, Alexious. 1972. *On Emotional Presentation.* Evanston, Illinois: Northwestern University Press.

Moore, G. E. 1903. *Principia Ethica.* Cambridge: Cambridge University Press.

———. 1922. "The Conception of Intrinsic Value" in *Philosophical Studies.* London: Routledge & Kegan Paul. pp. 253–75.

Naess, Arne. 1973. "The Shallow and the Deep, Long-Range Ecology Movement: A Summary." *Inquiry* 16: 95–100.

———. 1984. "Intuition, Intrinsic Value, and Deep Ecology." *The Ecologist* 14: 201-03.

Nagel, Thomas. 1980. "The Limits of Objectivity," in Sterling McMurrin, ed., *The Tanner Lectures on Human Values,* vol. 1. Salt Lake City: University of Utah Press. pp. 77–139.

Passmore, John. 1974. *Man's Responsibility for Nature.* London: Duckworth.

Rawls, John. 1971. *A Theory of Justice.* Cambridge: Harvard University Press.

Rolston, Holmes, III. 1974. "Is There an Ecological Ethic?" *Ethics* 85: 93–109.

———. 1987. *Environmental Ethics: Duties to and Values in the Natural World.* Philadelphia: Temple University.

Ross, W. D. 1930. *The Right and the Good.* Oxford: Clarendon.

Routley, R., and Routley, V. 1979. "Against the Inevitability of Human Chauvinism" in *Ethics and Problems of the 21st Century* ed. Goodpaster and Sayre. Notre Dame, Indiana: University of Notre Dame Press. pp. 36–59.

Scheler, Max. 1973. *Formalism in Ethics and Non-Formal Ethics of Value.* Evanston, Illinois: Northwestern University Press.

Taylor, Paul W. 1986. *Respect for Nature: A Theory of Environmental Ethics.* Princeton: Princeton University Press.

Wiggins, David. 1976. "Truth, Invention and the Meaning of Life." *Proceedings of the British Academy* 62: 331–78.

5

Environmental Values and Democratic Institutions

Christopher J. Bosso

In October 1990, Congress passed and President George Bush signed into law sweeping amendments to the 1970 Clean Air Act. The revised law, designed to address perceived deficiencies in federal regulatory policy, committed the nation to dramatic improvements in air quality by requiring (1) more stringent standards on automobile emissions and a wide array of industrial pollutants, (2) reduced use of high-sulfur coal in and more advanced pollution control by midwestern power plants, (3) sharply lowered air emissions of some 189 toxic substances, and (4) the eventual abandonment of ozone-depleting chlorofluorocarbons. The various new standards were to be phased in over a ten-year period, beginning in 1995, to allow affected industries and states time to comply.

Passage of the act came after more than a decade of deadlock within Congress and, during the 1980s, between congressional Democrats and a Reagan White House that had evidenced little, if any, enthusiasm for a more vigorous environmental agenda. Two factors appeared to dislodge this gridlock. First, in 1989 President Bush (who had campaigned in 1988 on a platform that promised renewed White House sensitivity to the environment) proposed his own revisions to the Clean Air Act. Although the final legislation went beyond the president's proposal—indeed, there was concern that Bush might oppose the final version as overly ambitious and costly—the administration's bill did spur congressional Democrats to put aside their internal warfare and negotiate seriously, if only to avoid the specter of being outflanked on the environment by a Republican president. Second, the Clean Air Act debate not coincidentally gained momentum about the time that the nation observed the twentieth anniversary of Earth Day (April 22), which seemed to reignite mass consciousness about the environment in general. With public attention to the environment so apparently heightened, and with congressional elections not far away, Congress and the president were loath to fail once again on a landmark piece of legislation.

Yet, only one month after the approval of the Clean Air Act, California voters convincingly defeated Proposition 128, a 16,000-word ballot question that had seemed a cinch to pass overwhelmingly during the previous summer. If enacted, Proposition 128 (known more popularly as "Big Green") would have marked the most dramatic and comprehensive attack on environmental problems taken by any American government entity, federal or state. Its proponents claimed that with the passage of Big Green, by the year 2000 California would have cut its carbon dioxide emissions by 20 percent, banned nineteen pesticides (by 1996), added new regulations for pesticide residues on food, embargoed off-shore oil drilling, placed state import taxes on oil, imposed stricter regulations on sewage treatment plants, outlawed clearcutting of redwoods, and banned all statewide use of chlorofluorocarbons (by 1997). The plan's supporters argued that only by bypassing a state legislature that many perceived as captive to powerful business interests could citizens impose their will directly and mount a comprehensive attack on the state's environmental problems. Californians overwhelming seemed to agree that their state's environmental woes required dramatic answers but, by November 1990, fears of war with Iraq and of domestic economic recession overcame previously solid public support for stronger, if more expensive, environmental protection. William Solesbury has noted that "political systems can cope with only a limited number of recognized issues at one time, . . . and these are always subject to displacement by new emerging issues of greater appeal and force" (1976, 13). Proposition 128 apparently was a victim of timing.

However, to chalk up Big Green's demise simply to bad luck is to ignore more fundamental factors that made passage more problematic than early public support might have led one to believe. It would be equally simplistic to assert that the labyrinthian battle over the Clean Air Act was due solely to the intransigence of the Reagan administration. Looked at more broadly, despite their apparent uniqueness both are telling illustrations of the American style of policy formation, processes by which problems are (or are not) brought to government and are (or are not) remedied. Notice the qualifiers. Nothing is at all certain about outcomes, because so much of policy making is about changing contexts, momentary opportunities, and shifting alliances. But it is not pure serendipity: policy making also takes place within the confines of a specific system of decision making, one whose particular tendencies make the prospects for comprehensive and effective environmental policy highly problematic, if not impossible.

This chapter examines the fundamental structure of the American system of policy making, a system based on the Constitution of 1787, as the key variable in understanding why it so hard to achieve comprehensive environmental policy in the United States. Other variables obviously come into play and in some cases are critical to success or failure. The impact of timing was evident in both cases mentioned above, yet to assume that timing alone determines policy outcomes is rather reductionist: after all, how does one know beforehand when the time is right? Political culture also obviously matters: a highly heterogenous and individualistic culture in which citizens are strongly skeptical of government (and of all institutions, for that matter) and highly protective of personal rights is not greatly suited to acting in support of governmental regulation. The United States is neither Sweden nor Japan. As David Vogel points out, "The relative formalism of the American regulatory system (its reliance on clearly defined rules and standards) reflects the inadequacy of informal mechanisms of social control within a highly individualistic culture." (Vogel 1986, 250). But culture is not destiny, because that same American public also at times accepts the need to cede (what they define as individual freedom) for the common good.

All other factors (timing, culture, technological change, the beliefs of individual policy makers, and so forth) must be discussed within the context of the formal system of government decision making. This chapter focuses narrowly on the structural design of the American system and its particular impacts on policy making because, simply put, institutions (and institutional designs) matter (Dolbeare 1985, 3). It may well be that the structure of the American system is antithetical to the sort of values and policy approaches central to contemporary environmentalism. This observation might seem revolutionary in light of the long-exerted hold of the behavioral social sciences over policy analysis, but it is, in a sense, a return to an older notion, one to which the framers of the Constitution certainly adhered.

The following section offers an overview of how the fundamental structure and processes embedded in the system of governance conditions the style of public policy making in the United States. It is, in many respects, an overview of the basics: how the design of the system itself influences the making of public policy and why, when all is said and done, the system is the problem if one desires comprehensive and integrative policy outputs. The next section assesses the values underpinning this system of governance and shows how they are in fundamental conflict with those that reside at the core of contemporary environmentalism. Finally, in light of all the apparent contradictions, we will examine why it is that effective environmental policy actually can get made (although the definition of "effectiveness" may differ from that propounded by economists or more quantitative policy analysts) and how those seeking policy change should take advantage of the very system that so often seems to impede their efforts. If the system is the problem, it also contains the elements whereby real, lasting, and, most important, legitimate environmental progress can be made.

The Formal Structure of Bias

Public policies do not emerge out of some vacuum but instead are crafted in an arena for decision making that has its own rules, procedures, and routines. More than three decades ago, E. E. Schattschneider argued simply that "organization is the mobilization of bias," that all forms of organization, whether as superficially simple as a local school board or as obviously complicated as the U.S. Congress, have embedded within them arrays of rules and procedures that, in the end, create arenas for decision making within which some types of interests and issues inevitably are favored over others (1960, 71). Schattschneider in no way argued that this bias resulted from crude ruling class dominance (as a simplistic Marxian analysis might lead one to suggest) but that *any* form of government, *any* institution, *any* law is in the end a form of bias. There is no such thing as a neutral law, nor any form of government (or market for that matter) where all interests are equally represented and influential. In the end, all forms of government, and all public policy processes, almost consistently favor some types of interests and some types of policy outcomes over others. It cannot be any other way.

Schattschneider's dictum is to be heeded here, because in their tendency to depict policy makers as devils many environmental activists forget that policies are knitted together within structures of decision making that have identifiable biases, many of them in direct conflict with the types of thinking that underpin ecological values. If politics is the art of the possible, the formal system in which politics takes place affects what *is* possible. In the U.S. context, the fundamental bias is the Constitution itself, based as it is on the seventeenth-century Enlightenment premise that humans are rational beings who act according to their respective self-interests. Each individual is assumed to be rather self-sufficient, so extensive governmental intervention is unnecessary. However, because humans are also flawed, citizens of a polity recognize the need for a framework of order and stability to stave off the possibility that some may abuse the inalienable rights of others, including that to private property. Citizens thus join together in a social contract to create a government that will allow them, within limits, to pursue freely their separate interests.

This conception of government as a contractual arrangement is highly *mechanistic*. That is, rational actors create a mechanism of laws (government) that will condition how all are to behave in civil society and, equally important, how their government is to treat them. As James Madison so pungently argues in *Federalist* #51, "In framing a government which is to be administered by men over men, the great difficulty lies in this: you must first enable the government to control the governed; and, in the next place, oblige it to control itself." (Hamilton, et al 1961, 322) This pervasive wariness of government, when paired with an equally palpable fear that any single "faction" might usurp all governing power, is manifest in a Constitution that purposely fragments such power and diffuses authority to deter unfettered dominance by a few elites, but which also undercuts the capacity of possibly antidemocratic majorities to have their way. It is *not* majority government. As Dahl and Lindblom write, "The strategic consequences of this [constitutional] arrangement . . . has been that *no unified, cohesive, acknowledged, and legitimate representative-leaders of the 'national majority' exist in the United States*"(1953, 336).

This "Madisonian" constitutional device is thus a systemic *centrifuge* designed to allow as many as possible to pursue freely their particular (and probably localized) interests without enabling any one faction, even a majority, to wield power totally or quickly. In institutionalizing conflict, this mechanism seeks an equilibrium among the vast array of societal forces to which it nonetheless grants varying degrees of access: stability and incremental change are innate tendencies in a system that forces coalition building bloc by bloc, parochial interest by parochial interest. It is, in many ways, a bottom-up process by which "majorities of the moment" are knitted together out of a great many diverse interests. More pertinent to our discussion, this system was designed first and foremost to deter elite and mass tyranny, *not* to enable an activist government to plan and implement comprehensive national policies, a failing that emerges most forcefully when problem realms are ill structured and diffuse or where innovative answers are needed. The American system was designed to do little beyond safeguarding the union and the rights of its citizens: nowhere in the original scheme were the mechanisms necessary for national policy making present, beyond that which could be cobbled together in the legislature.

Federalism

This systemic centrifuge first reveals itself in the allocation of governing responsibilities between the federal government and the various states. Federalism in many ways makes the United States possible, because spinning some aspects of governance out to the states both allows policy makers to address local needs more easily and reduces aggregate pressures on the central government. It is often forgotten, in this era of federal preemption of local prerogatives, that the states continue to have significant governing authority distinct from any enjoyed or wielded by the national government. The limited autonomy of the states, by itself, makes policy making more Byzantine as compared to a more unitary system (e.g., France), where all policy decisions (as opposed to the merely administrative) are made by the central government. The picture in the American context is made all the more variegated by this nation's sheer vastness, its mind-boggling physical diversity, and, unforeseen by the founders, its unprecedented demographic heterogeneity. "The economico-political diversities which government has to harmonize in the United States are thus not only more numerous," writes the English scholar H. G. Nicholas, "they are also more distinct and exclusive"(1980, 3). After all, the array of problems exhibited by Los Angeles are probably not shared by Helena, Montana, and widely diverse state traits, resources, and political inclinations are certain recipes for a numbing patchwork of laws and policy directions.

Because they exist, and because they have power, the states are singly and collectively the Achilles' heel of effective national policy. Comprehensive policies, where the whole may in fact be more overarching more often than not, fall victim to local and regional needs as refracted in Congress, and particularly in that vestige of the principle of state sovereignty, the Senate. Forging national policy in the United States is usually less a matter of divining some putative common good than patching together a thin quilt of compromises, one that covers parts of the nation less adequately than needed but which certainly is far more than other states would have provided for themselves. National policies also are likelier than not to be highly legalistic, dependent on complicated formulas and exacting standards that attempt to take into account the scope of state and regional differences, but which also cannot help but produce greater expense and delay than probably results in smaller and more unitary polities.

Federalism, and the leverage wielded by state interests in Congress, also makes it very difficult for the national government to impose any degree of localized pain for some broader national good. Avoiding for a moment the normative questions, it is obvious that the nation needs to somehow dispose of its toxic and radioactive wastes, but the federal government has been remarkably impotent in overcoming state resistance (just as the states themselves are blocked consistently by local opposition). In fact, one can argue that the success of *not in my backyard* (NIMBY) politics in the United States is aided and abetted by a system where the center has a very difficult time imposing anything and where there is a strong predisposition to respect the rights of each and every state. Not unless a single state can be shanghaied by the others (with the usual array of payoffs to lessen the inequity of it all) can such policies be carried out, and the state in question certainly has the opportunity to appeal the outcome in the federal courts. The result is more delay and the greater probability that the "solution" will hardly suffice.

On the bright side, the individual states are the chief laboratories for nurturing new ideas, some of which "diffuse" to other states or are adopted nationally (usually in some diluted form) by Congress. For example, California has consistently gone beyond the federal government in the stringency of its auto emissions standards (limits recently adopted by Massachusetts), while New Jersey proved an important incubator of ideas for handling toxic chemical wastes. But this role as policy vanguard is limited, because environmental problems are increasingly transboundary in nature and eventually require some degree of national government coordination or intervention. More critically, no state can be seen as too "radical" on environmental matters. States that pass tough industrial emissions laws may find themselves losing factories to those where the laws are less exacting, a fact of economic mobility in a free-market system that plays nicely into the hands of state business interests. This capacity to play one state against another takes on even greater impact when an individual state depends disproportionately on a single industry, (e.g., Louisiana and petrochemicals) for its very livelihood.

The advantages and weaknesses inherent in this system of vertical separation of powers were played out with California's Proposition 128. To wage such a comprehensive, multi-problem attack on environmental degradation would be unthinkable in Congress, for reasons to be addressed below. But the initiative system in California is a sublime example of how the diversity of state laws can provide for tremendous innovation. Yet, timing notwithstanding, the very size and audacity of Big Green was its undoing. So massive in scope was this plan that it affected virtually every industry in the state, and any ripple effect from California's immense market would resonate far beyond its borders. It was thus no surprise that the tally sheet of those state, national, and international interests that opposed some segment of Proposition 128 far outstripped its supporters. Given the ballot proposition's manner of presentation, the only way each particular interest could affect its portion of the initiative was to fight against the whole package. Furthermore, opponents of Big Green could persuasively point out that passage of the proposition would place California at a tremendous competitive disadvantage and that it would be foolhardy to try to address such an extensive array of highly intractable problems within the boundaries of a single state. Any way it added up, Big Green was simply too big for one state to swallow.

Separation of Powers

In the national government these mechanistic impulses are played out in the rationing of governing power among three distinct but not quite independent branches, each responsible for somewhat separate functions but all in the end sharing power so that none of the three dominates the others unchecked. Institutional conflict and jurisdictional rivalry is guaranteed, even absent partisan and ideological frictions, and a Constitution that is terse for the ages certainly does little to clarify how government is supposed to resolve contemporary problems. As Joel Aberbach suggests, it all was intentional: "The grand design of the system the framers devised was one in which every actor or institution would be checked by another. Power would be diffused, individuals and institutions would be forced to accommodate one another, and the ambition of political actors would be harnessed, the framers hoped, for the public good" (1990, 3).

The public good (or, at minimum, the momentary public mood) was to be ascertained through the election of representatives to the national legislature. In parliamentary democracies (e.g., Great Britain) the majority will is arguably refracted through the election of the governing party which, because it holds the majority of seats in the legislature, can govern more or less as it sees fit. What is more, parliamentary systems allow their governments to reaffirm their public support by calling for new elections when the omens seem propitious or when a failed government must reconstitute itself. The United States, however, biannually and without exception, engages in a process whereby all House members and one-third of all Senators submit themselves for reelection individually and separately before their own narrowly configured constituencies. Although refracting arguably accurate representation of local and state interests, this system is the source of fragmented policy making, not some embodiment of the majority will. If all members of Congress do their job (and reelection rates for incumbents offer at least superficial proof that they do), the cumulative result is a Congress of 535 free lancers, each responsive primarily to the folks back home. Voters, for their part, adhere well to former House Speaker Tip O'Neill's maxim that "all politics is local," at least insofar as they complain that Congress, as an institution, does not work, while in the same breath rating their own incumbents highly on constituent services and attention to local needs. Since the late 1940s, voters also have confused matters further by typically granting control over Congress to one party and the presidency to the other, as if to say that the formal separation of powers embedded in the Constitution is not enough of a check on government power.

Members of Congress, of course, understand these dynamics and never fail to cater to their constituents even if it means bucking their party or president, or both, on important national policy questions. The hoary strategy, "vote for every appropriation and against every tax increase," is one that works almost every time, particularly when there are so few sanctions against such fecklessness. One expects this parochial pandering in the House of Representatives, which, after all, only adheres to its constitutionally intended role as the barometer of the momentary passions of the public. But the democratization of Senate elections in the early twentieth century and their subsequent elevation to intensely media-driven contests since the early 1970s have produced in senators many of these same localistic traits. The congressional election system thus produces representatives of the many parts, but none of the whole.

Related to this systemic electoral centrifuge is the matter of the two-party system, although the relationship may not be obvious at first. However, the congressional election system's emphasis on winner-take-all contests and single-member constituencies works directly against the sorts of multi-party systems typical of parliamentary democracy. Small parties cannot win in the American system, because it takes at least a plurality of votes to win congressional seats and an absolute majority of the electoral votes to attain the presidency. The result is a dynamic of pre-election coalition building among a great many interests. The American political party is not the result of some societal cleavage, where members of the same caste or occupation are bound together by some common outlook, but instead is an amalgam of interests bonded together largely by a collective desire to win some slice of formal political power. Parliamentary democracies, with their many variations on proportional representation, generally enable smaller and more ideologically cohesive parties to win seats in the national legislature proportional to their electoral strength.

Coalition building in such systems takes place *after* the election, in the legislature, especially if no single party attains a majority sufficient to form a government. It comes as no surprise that "green" parties have won seats in legislatures in many European parliamentary systems, but no such party even exists nationally in the United States.

The electoral system is thus biased toward a two-party oligopoly, and environmentalists, like everyone else, must jostle for advantage within either of the two centrist behemoths, where their messages must compete openly with all sorts of other programs and objectives to even get atop a party agenda, which itself is worth little after the election passes. It therefore is no surprise that so few voters regard the parties as adequate vehicles for the transmission of particular issues to government or that so many have turned to more direct forms of participation, through organized interests, to represent their views. The pervasiveness of interest group politics in the United States is related directly to the weakness of its party system and to the relative ease with which non-party groups can mobilize and gain access to at least some part of decision making. Interest groups may embody the vibrancy of a free society, but they individually and collectively do not aggregate citizens into a "public"; they instead fragment the public further into more narrowly configured "factions" than even the framers of the Constitution might have imagined. When combined with the purposeful diffusion of governing power, this mélange of interests provides for countless narrow and unconnected policy debates but offers little to the needs of broader and more integrated governance. The picture one gets, to borrow an image from Anthony King, is an "atomistic" political landscape (1978, 391).

In sum, the election (and political party) system provided for by the Constitution does not enable the majority will to be known unless one presumes (wrongly) that the sum of many and separate interests adds up to a majority will. The American election system simply provides for the selection of unattached individuals for identifiable and equally unattached seats in government, each parceled out for the representation of some narrow geographic constituency. The result is less a Congress of two parties than a collection of individual legislative entrepreneurs, each servicing an exclusive set of customers. What elections do not produce is a *government*. This is no surprise, given the framers' straightforward fear of majority tyranny.

Neither does one find in the American electoral system the flexibility for calling elections to reaffirm the performance of the government or to reconstitute one that has failed: the American mechanism operates by a programmed clock, and the practical horizon for elected officials extends no farther than the next campaign (Sundquist 1986).

Furthermore, and reflecting legislators' own electoral and policy needs, the centrifuge spun by the election system has become institutionalized within congressional structures, processes and norms. The offices of individual members are today little more than constituency service providers, shops oriented to the member's own reelection. Party loyalty, the glue that once arguably provided some meager counterweight to the system's innate centrifugal tendencies, in most instances no longer stands up to voter demands and the siren song of organized interests, which themselves focus on narrow slices of public policy. Still-evident institutional norms of reciprocity, mutual non-interference and respect for the rights of all members (especially in the hyperindividualistic Senate) make for an institution where representation of narrow constituency interests takes precedence over broad national initiatives. As Charles O. Jones suggests, "The net result for the policy process appears to be an institution with many more claims but not necessarily the means for consolidating these claims into coherent policy proposals" (1984, 11).

If individual members of Congress are the entrepreneurs, the House and Senate standing committees and subcommittees are their shops. On paper the committee system looks as if it were designed along efficiency-seeking functional lines (agriculture, education, commerce, and so on), and it arguably allows for a division of labor and the fostering of policy expertise, which, in turn, theoretically enhances the capacity of Congress as an organization to do its work. But committees are also the primary platforms from which members protect constituent interests and pursue personal policy goals, and the positions of a well-placed few all too frequently allow them to wield vetoes over public policy. Committee rosters not surprisingly are stacked with those members who have the closest possible attachments to prevailing policies and appropriations, and their defense of the status quo is augmented by their cozy relationships with those executive branch bureaucrats who administer these programs and those organized interests dependent on committee protection. To offer but one example, the House Agriculture Committee's long-held jurisdiction over the regulation of chemical pesticides gives farm and chemical industry interests a tremendous advantage over environmental groups each time demands for wholesale policy change arise. Environmentalists may rail against committee members as mere defenders of some unacceptable ecological status quo, but their pleas more often than not are drowned out by the siren songs of constituents and their affiliated organized interests (Bosso 1987). This institutionally buttressed ability of narrow interests to screen out contrary views is replicated throughout the congressional committee system.

To make matters more complicated, it is not all that uncommon to find several committees fighting with one another for jurisdiction over key policy areas. For example, and to the dismay of the Agriculture Committee, nearly a dozen House and Senate committees and subcommittees have at one time or another staked out at least partial jurisdiction over some aspect of federal pesticide policy (Bosso 1987, 188–89). Although this competition offers environmentalists avenues for access, it also creates turf wars that make reaching agreement all the more problematic. More broadly, at least seven House and five Senate standing committees (and innumerable subcommittees) wield control over some slice of overall federal environmental policy (Vig and Kraft 1990, 6). This fragmentation and segmentation may have sufficed when issues arguably were more discrete (e.g., agriculture did not directly affect the merchant marine) and where programs could be distributive in nature (e.g., public works projects), but it has become entirely dysfunctional in issue realms that transcend committee boundaries and that require some element of coordinated policy making. As Michael Kraft argues, "The consequences of such jurisdictional complexity include delay in addressing some policy needs, piecemeal consideration of related policies handled by different committees, and inconsistencies and lack of coherence in the nation's environmental policy" (Kraft 1990, 191).

Congress as an institution has multiple functions (lawmaking, representation, executive oversight, and so on) but there is little doubt that since the late 1960s representation has been the overriding obsession, with members all too ready to cater to the folks back home regardless of the impacts on national policy. Congress, since the early 1970s, has also become a much more open institution, with power more dispersed, thus transforming the once textbook picture of Congress as a pyramid of influence (with the committee chairs at the apex) into an image almost as inchoate as a Jackson Pollock painting. Order is replaced by disorder, party legions by entrepreneurial individualists, committee government by subcommittee warfare, and all is much more accessible to outside influences, whether they be the media, constituents, organized interests, or the president. The congressional system, if one can call it that, is one in which the access points have become many and the veto points endless. It is a permeable institution, with access almost guaranteed. Representation in the sense that Edmund Burke might have imagined (where legislators act as trustees of some broader public good) is replaced here by catering to narrow organized interests, each of which seems capable, within this arena of segmented power, of stopping policy initiatives in their tracks.

Major legislative breakthroughs do occur, but as the passage of the 1990 Clean Air Act attests, the path to nonincremental policy change is an especially tortuous one. In the House, Representative John Dingell (D–MI) used his position as chair of the Energy and Commerce Committee (through which almost all air pollution—related legislation had to travel) to protect the interests of domestic auto manufacturers located in his state by bottling up bills mandating more stringent auto emissions. For years in the Senate (bicameralism complicates matters further), Majority Leader Robert Byrd (D–WV) actively thwarted floor debate on acid rain legislation for fear that cutting electric utility use of high-sulfur coal would cost thousands of West Virginia coal miners their jobs. Not until Senator George Mitchell (D–ME)(whose state was a major acid rain victim) became majority leader did acid rain legislation enjoy a more privileged position on the Senate calendar. The menu of local and regional obstacles to major statutory change went beyond these few individuals in key positions: westerners strongly opposed stricter air quality standards for their sparsely populated states; members from states bordering the Gulf of Mexico resisted tighter standards on emissions from off-shore oil wells; farm state representatives feared impacts on agricultural practices.

As might be expected, the Clean Air Act was hardly the product of careful planning. It came about instead through a long string of compromises, during which a colorful mosaic of parochial concerns was pieced together until it was broad enough to garner a working majority of support for the final bill. In fact, so many and varied were the affected interests (and so successfully did each and every threatened interest impede action) that not until President Bush proposed his own bill were opponents forced to sit down and negotiate. This process was aided greatly by the usual bargaining, all of which, critics argue, probably will serve eventually to undermine the law's aggressive goals.

The acid rain agreements were the toughest to negotiate, because they pitted "clean" states (those that did not burn high-sulfur coal) against "dirty" states, mostly in the Midwest. At the crux of the struggle was the question of how to distribute the costs of stricter emissions control: the "clean" states wanted to make the "dirty" states pay the entire cost of bringing their power plants into compliance, because those states had been the most recalcitrant in the past. Those states, not surprisingly, wanted to spread out costs nationally. Negotiators naturally avoided "zero-sum" outcomes (which the system does not impose well anyway) and came up with a complicated system of "pollution allowances" for industries and utilities. Midwestern states were coaxed along with generous extra allowances, a concession environmentalists opposed, even though the future costs to those states will still be high.

Other deals included special rules for coke plants, essential to the production of steel; new assistance programs for displaced workers, especially coal miners; exemptions for Texas, Louisiana, Mississippi and Alabama from stricter controls on off-shore oil platforms; exemptions for most farm equipment; separate consideration on toxic emissions for research facilities; special exemptions for Florida and for a power plant in Michigan; special grants for studies on acid rain at the University of Wyoming and on liquefied natural gas at the Universities of Nevada and Wyoming; exemptions for emissions in the production of, and equipment used in the exploration, production, development, storage and processing of, oil from "stripper" wells; a multi-year study of the effects of acid rain on the Adirondacks to be conducted only at the Rensselaer Polytechnic Institute; and set-asides on EPA-funded research for disadvantaged business concerns (*Congressional Quarterly Weekly Report*, October 27, 1990, 3587).

What finally emerged aptly underlines Walter Rosenbaum's argument that "extensive negotiation among many participants [the price of securing support for major policy in a fragmented power system] means that the policy that emerges is often the common denominator of many different preferences" (1973, 99). The lessons of the 1970 Clean Air Act, a package of laws that Charles O. Jones labeled "policy beyond capabilities," leaves one to wonder whether the "common denominator" amendments passed in 1990 will actually solve the problem (Jones 1975).

Note here that the Clean Air Act probably never would have come to pass were it not for the intervention by the president, whose institutional capacity to set the agenda of government and mobilize support for favored initiatives knows no parallel in the American system. One expects Congress to reflect the largely parochial and short-term interests of constituents: that, after all, is its ostensible purpose. The presidency, on the other hand, was intended by the framers to represent the nation as a whole, to offer a view counter to that promoted within Congress. The presidency (and the executive establishment generally) theoretically offers the best hope within the federal government for truly sensitive ecological thinking and planning, because the president alone among elected officials has the arguable mandate to integrate national policies and the institutional capacity to think in broad and relatively long-term ways. This institutional role is all the more critical in this era of divided partisan control of the branches, because the president usually cannot expect undiluted party support. The courts, for their part, decide only segments of policy, confined as they are to real cases and constrained from sweeping pronouncements in all but the most exceptional situations by precedent, legislative intent, and tradition. In a sense, only the executive branch has the institutional capacity necessary for effective national environmental policy making.

However, American presidents, too, are creatures of the constitutional centrifuge, constrained in domestic policy in more ways than their prime ministerial counterparts can imagine and operating in an adversarial arena that offers remarkably little continuity. The presidency, as an institution, is no less fragmented internally, driven by short-term political goals and pulled in multiple directions by competing policy objectives. The environment must in fact compete for the president's attention, and there is no doubt that its relatively low priority on the Reagan agenda during the 1980s contributed mightily to the subsequent gridlock across the board. President Bush, in the heat of the electoral battle, proclaimed his fealty to environmental progress, but the coming threats of war with Iraq, concerns about the faltering domestic economy and tussles with Congress over the staggering federal budget deficit quickly yanked the environment off the president's set of priorities, to be left to the mercy of normal congressional and bureaucratic politics.

This personalization of presidential policy making would not matter so much if the federal administrative sector enjoyed the independence, power and prestige of its European and Japanese cousins. But it does not. In fact, despite the popular image of rampant bureaucratic government, the federal bureaucracy is remarkably fragmented internally and constrained externally by Congress, the courts and, in many cases, the mass media. These institutional realities are buttressed by (if not the creations of) a culture that despises government and is loath to grant either power or prestige to unelected officialdom. On paper (but only on paper) the Environmental Protection Agency has primacy over environmental management, but it is hardly independent of the White House (whose budgetary and partisan goals may not agree with the EPA's mandate) and must battle continuously with the other departments and agencies that wield control over segments of national environmental policy. To make matters worse for the EPA, its bureaucratic competitors usually have congressional patrons and interest group allies who are far more focused and determined than ever seems the case for the broad and rather fragmented environmental community. To put the matter bluntly, the EPA is outgunned: it does not even have a guaranteed place in the cabinet, a seemingly minor matter that nonetheless places the agency at a disadvantage relative to such departments as Agriculture and Energy. President Bush in 1989 promised to elevate the agency to departmental status, but the effort bogged down in Congress as shifting priorities and bureaucratic warfare eroded the momentum.

A short mention should be made here about the judiciary, which in the past three decades has come to play a major and often unforeseen role in environmental policy formation and, perhaps more important, implementation. In American common law, individual property rights traditionally have taken precedence over some broader community good, and corporations as entities traditionally were viewed as equivalent to private individuals. Additionally, long-held norms required that damage suits were to be based on tangible proof of direct harm, a standard harder to meet than environmentalists might care to admit. The combination of these biases made the courts throughout history tend to defend the rights of the individual citizen or corporation, even when an individual's actions had spill-over effects. Even the opening up of the court system in the early 1960s has not erased this bias. In addition, the great silences of the Constitution and the inability of Congress to specify every technical detail in every environmental statute creates for the courts challenges that are usually met in a segmented and narrowly drawn fashion, typically ordering a halt to some practice or enforcing standards on the books. The courts cannot be expected to act otherwise, for they lack the institutional capacities and political legitimacy to plan, and the cumulation of discrete precedents hardly equals an integrated national environmental policy.

Conclusion

Almost twenty years ago, Walter Rosenbaum, in scanning the American political system and its response to the environmental wave of the early 1970s, argued that "the environmental movement places a set of demands on Washington policy makers that the policy process has so far not been well adapted to meet." (1973, 103). He goes on to state:

> Achieving a comprehensive governmental approach to environmental management is a formidable task, for environmental policies have been neither comprehensively planned nor controlled. Rather, planning and authority have been fragmented into a bewildering mosaic of different agencies and levels of government, producing frequent confusion, contradiction, and frustration in environmental management. (1973, 108)

The trends witnessed over the past two decades have not simplified matters. In fact, one can argue that the policy-making system as a whole has become even more open, more fragmented, and more adversarial. As Vig and Kraft argue twenty years later:

> The effect of divided authority is slow and incremental alterations in policy, typically after broad consultation and agreement among diverse interests both within and outside of government. Such political interaction and accommodation of interests enhance the overall legitimacy of the public policies that result, but over time the cumulative effect for the environment has been disjointed policies that fall short of ecological or holistic principles of policy design. (1990, 8)

More daunting, the types of environmental problems that the system could not handle twenty years ago have been superseded by successive offspring that are increasingly transboundary and nondivisible in nature. First generation environmental problems (for example, conserving parklands or cleaning up localized water pollution) in some ways could be addressed in the time-honored American fashion of distributive policy making; that is, handing out federal grants for water pollution control, new parks, and so on. Almost everybody will want the free money, and as long as the funds are not seen as coming at the expense of somebody else (i.e., a positive-sum solution), few will fight much about it. But newer types of environmental problems are not so amenable to these traditional distributive answers. The problems created by agricultural groundwater contamination, much less global warming, cannot be solved merely by allocating monies to the various states and congressional districts. They require long-term planning and multi-jurisdictional coordination. In short, they may require the types of administrative and governing capacities that simply do not reside in the American system.

Ecological Values Versus Adversarial Democracy

David Vogel, in comparing British and American styles of environmental regulation, suggests that "the characteristics of a political regime are more important than the nature of the particular policy area itself in explaining policy processes" (1986, 195). Vogel certainly is correct insofar as policy making is not *determined* (as opposed to affected) by policy types. But, even as we argue that the structure of the American system itself is a bane to comprehensive policy making, it is necessary to add that some aspects of environmentalism *as a value domain* do seem to conflict more sharply with these fundamental institutional biases than do other policy issues (e.g., social security). As Walter Rosenbaum argues, "Despite the great profusion and complexity of decision-making procedures in American government, in the United States (as in other governmental systems) the rules tend to correspond with certain principles or values that are considered essential to the governmental process" (1973, 94). If environmental policy has been disjointed, parochial, and short-term, perhaps the single reason is the vast disparity between the values underpinning American constitutional government and those at the heart of contemporary environmentalism.

Table 5.1 lays out a few of the more obvious (and probably simplistic) contrasts between the values that make up the foundation of the American system and those central to contemporary environmentalism. The first, and probably most important, difference is the "science" that produced each value system. The American style of governance emerged during the heyday of Newtonian science, a period in which "mechanistic" conceptions permeated all aspects of thought, including the political. Eighteenth century constitutional democracy thus emerged as a "process" model

Table 5.1 System and Environment Values

System	*Environmental*
Mechanistic	Organic
Segmentation	Interconnectedness
Individual freedom	Common good
Short term	Long term
Localism	Globalism
Antiplanning	Proplanning
Adversarial	Cooperative
Incrementalist	Comprehensiveness

based on the belief that the proper mechanism was the basis for effective and accountable government. More important, this machine was to be self-regulating; it was meant as a thermostat of sorts that would maintain social stability and prevent the abuse of rights even if its officials were not the most talented or even ethical. In the American context, this view was colored by the founders' pervasive fear of power and their desire to "make ambition counteract ambition." The result is adversarial democracy, a system of representation and governance that institutionalizes conflict and fractures policy making into narrow segments to prevent any one group from holding all political power.

Environmental values seem to transcend particular processes and are in the main based on organic or biological notions of the relationship between humanity and nature, assumptions that simply are alien to the philosophical foundations of eighteenth century mechanistic thought and, perhaps equally important, are contrary to powerful Judeo-Christian traditions stressing the dominion of Man over Nature. Ecological thinking stresses the interconnectedness of the parts that make up the whole and that each part in its place is essential to the overall survival of the ecosystem. These parts are *not* interchangeable: the loss of one affects the whole in subtle and potentially devastating ways. To many environmentalists, specific governmental processes really do not matter (theoretically, a dictatorship could be environmentally sensitive) so long as these processes stress the common good and avoid conflicts that could undermine the organic whole. Although some theorists argue strongly that environmentalism is not contradictory to progressive democratic values (Paehlke 1989), one has to admit reluctantly that democracy in and of itself is not a necessary and sufficient condition for effective and "efficient" (in the narrowly economic sense) environmental policy making.

The American system emphasizes the rights of the individual, including the ever-critical right to private property. Environmentalism at its core stresses community: shared needs, public goods, the need to perhaps subordinate some facet of the individual to the collectivity. These two angles of vision are not inevitably hostile, but they certainly are not made easily compatible in the warp and woof of everyday life, especially taking into account the added privatizing force generated by a keen American adherence to rather robust free-market principles. Individualism is woven throughout the system, and it fragments the whole into many discrete parts. Any value system that stresses the needs of the whole over the freedom of individuals to pursue their own separate interests is typically rejected as "un-American."

The American constitutional mechanism also operates according to fixed time periods, most of rather short duration, because the need to return elected officials regularly and predictably back to the voters was the key to keeping government accountable and responsive to identifiable (and largely local) constituents. However, the founders did not want this barometer of the public passions totally determining public policy, so they also staggered the length of terms so as to make it harder for any one faction to attain power too quickly, or for long. True change would have to come gradually, not in sweeping majorities of the moment, because to allow for unchecked majoritarianism was to take the chance that the rights of the minority faction would be abused. The electoral system's foreshortened horizon disrupts any policy making that requires tremendous amounts of information and an institutional capacity to plan into the future. The system is anti-planning to its very marrow, but effective environmental policy may require just such planning.

The system also values localism, devolving as it does public power and underlining the societal predisposition to individual self-reliance. Yet contemporary environmentalism increasingly stresses more "global" values, and one wonders whether the slogan Think Globally, Act Locally makes analytical sense. It may in practical terms (even cleaning up local emissions is good for the world environment), but the rhetorical power of the global perspective may simply move activists toward proposals whose scope is beyond the capacity of the system to handle. California's Proposition 128 came to its ignominious end perhaps because its supporters were thinking too globally and acting too locally. At the risk of sounding overly harsh, the activists in this case were presumptuous about their own capacities and about the willingness of voters to unreservedly swallow a huge package of dramatic changes whose future costs and impacts were debatable at best, entirely unforeseeable at worst.

Most of all, effective environmental policy making requires activist government, on whatever level, but this very idea is hobbled in the American context by a framework of checks and balances and by a systemic predisposition toward limited government that makes such activism problematic to begin with and too often ham-handed when attempted. Simply put, the system is a conservative one, in the sense that movement is a democratizing force, one that gives average citizens a real say in the policies that affect their daily lives. In any form of democracy, the most "effective" policy is that which is legitimate in the eyes of the people, one that they will accept and defend. Too often in the past environmentalists were perceived as elitists and thus were dismissed by many as little more than another "special interest." Too many environmental groups have entered the 1990s as little more than "checkbook" organizations, granting a minimal role (if any) to members (Bosso 1990). It thus is no surprise that the most energized part of the environmental movement is that part where citizens can have the most immediate impacts on problems that affect their everyday lives. If environmentalists reach out beyond their traditional constituencies (particularly to the working class, the poor and minorities), they will indeed accelerate the process of reconstituting the public value system even as their sheer numbers give the movement as a whole greater clout in more immediate political battles. To deride grass roots activism simply as rampant and unenlightened NIMBY-ism is to ignore the very real democratizing potential these groups represent. If average citizens educate themselves to defend their neighborhoods and homes against perceived hazards, they might in time look out beyond their localized horizons to less immediate but equally important matters. Perhaps that slogan, Think Globally, Act Locally, isn't so contradictory after all.

Finally, much of the current apprehension about the perceived pathologies of the American form of government seems focused on its tendency to become deadlocked when conflict "gets out of hand." Scholars warn darkly about "system overload" or "hyperpluralism," as if to argue that broader societal participation is to be avoided, that the system itself cannot handle the burden. However, it seems upon reflection that this deadlock is the outgrowth of a healthier equilibrium among societal interests. "The pressure system makes sense only as the political instrument of a segment of the community," argued Schattschneider. "It gets results by being selective and biased; *if everybody got into the act the unique advantage of this form of organization would be destroyed, for it is possible that if all interests could be mobilized the result would be stalemate*" (1960, 34–35 [emphasis in original]). The current deadlock reflects the system's inability to force closure on issues upon which there is no solid consensus, especially in times when the society itself is making the often difficult transition from one set of underlying values to another.

American society may well be in such a transition, with the deadlocks of the 1980s reflecting the struggle between an old and the new sets of values. If this is the case, the current deadlock may well dissipate when the new values take hold. Environmentalists must take advantage of the opportunity to help shape those values. Their venues for such change, as the founders intended, are placed firmly among the people.

References

Aberbach, Joel. 1990. *Keeping a Watchful Eye: The Politics of Congressional Oversight.* Washington, DC: The Brookings Institution.

Borelli, Peter. 1987. "Environmentalism at the Crossroads." *Amicus Journal* 9: 24–37.

Bosso, Christopher J. 1987. *Pesticides and Politics: The Life Cycle of a Public Issue.* Pittsburgh, PA: University of Pittsburgh Press.

———. 1990. "Adaptation and Change in the Environmental Movement," in

Allan J. Cigler and Burdett A. Loomis, eds., *Interest Group Politics,* 3d ed. Washington, DC: Congressional Quarterly Press.

Browne, William P. 1989. *Private Interests, Public Policy, and American Agriculture.* Lawrence, KS: University Press of Kansas.

Dahl, Robert A., and Charles E. Lindblom. 1953. *Politics, Economics and Welfare.* New York: Harper & Row.

———. 1956. *A Preface to Democratic Theory.* Chicago: University of Chicago Press.

Davidson, Roger H. 1981. "Subcommittee Government: New Channels for Policy Making," in Thomas Mann and Norman J. Ornstein, eds., *The New Congress.* Washington, DC: American Enterprise Institute: 99–133.

Dolbeare, Kenneth. 1985. *Democracy at Risk: The Politics of Economic Renewal.* New York: Chatham House.

Hamilton, Alexander, James Madison, and John Jay. 1961. *The Federalist Papers,* Clinton Rossiter, ed. New York: New American Library.

Hays, Samuel P. 1987. *Beauty, Health and Permanence: Environmental Politics in the United States,1955–1985.* New York: Oxford University Press.

Jones, Charles O. 1975. *Clean Air: The Policies and Politics of Pollution Control.* Pittsburgh, PA: University of Pittsburgh Press.

———. 1984. *An Introduction to the Study of Public Policy,* 3d ed. Monterey, CA: Brooks/Cole.

King, Anthony. 1978. "The American Polity in the Late 1970s: Building Coalitions in the Sand," in Anthony King, ed., *The New American Political System.* Washington, DC: The American Enterprise Institute: 371–96.

Kraft, Michael E. 1990. "Congress and Environmental Policy," in James P. Lester, ed., *Environmental Politics and Policy: Theory and Evidence.* Durham, NC: Duke University Press.

Lindblom, Charles E. 1977. *Politics and Markets: The World's Political-Economic Systems.* New York: Basic Books.

Loomis, Burdett A. 1988. *The New American Politician: Ambition, Entrepreneurship, and the Changing Face of Political Life.* New York: Basic Books.

Lundqvist, Lennart J. 1980. *The Hare and the Tortoise: Clean Air Policies in the United States and Sweden.* Ann Arbor, MI: University of Michigan Press.

Mansbridge, Jane J. 1980. *Beyond Adversary Democracy.* New York: Basic Books.

Nicholas, H. G. 1980. *The Nature of American Politics.* New York: Oxford University Press.

Paehlke, Robert. 1989. *Environmentalism and the Future of Progressive Politics.* New Haven, CT: Yale University Press.

Rosenbaum, Walter A. 1973. *The Politics of Environmental Concern.* New York: Praeger.

Schattschneider, E. E. 1960. *The Semi-Sovereign People: A Realist's View of Democracy in America.* Hinsdale, IL: Dryden Press.

Solesbury, William. 1976. "Issues and Innovations in Britain, West Germany, and California." *Policy Analysis* 2 (Winter).

Steinhart, Peter. 1990. "Bridging the Gap: Can Earth Day 1990 Bring Together Greens and Mainline Conservationists?" *Audubon* (January): 20–23.

Sundquist, James. 1986. *Constitutional Reform and Effective Government.* Washington, DC: The Brookings Institution.

Vig, Norman J., and Michael E. Kraft, eds. 1984. *Environmental Policy in the 1980s: Reagan's New Agenda.* Washington, DC: Congressional Quarterly Press.

———. 1990. *Environmental Policy in the 1990s.* Washington, DC: Congressional Quarterly Press.

Vogel, David. 1986. *National Styles of Regulation: Environmental Policy in Great Britain and the United States.* Ithaca, NY: Cornell University Press.

Wildavsky, Aaron. 1979. *Speaking Truth to Power: The Art and Craft of Policy Analysis.* Boston: Little, Brown.

6

Science, Environmental Values, and Policy Prescriptions

David E. Henderson

Science began as a way to free people from irrational fears about the nature of the physical world. It helped eliminate superstitions and provided a rational way of viewing the world, and in many ways it still does. We know what causes tornadoes and earthquakes, even though we cannot predict them reliably. Science and the technology that results from scientific discoveries have taken on a new power as they provide tools for improving our lives. Unfortunately, science now also brings us fear. Much of this fear is the result of our increased ability to measure natural and synthetic substances in our environment. Some of the fear is a result of our increasing understanding of the effects of human activities on the ecosystem.

Equally significant to the developing role of science is the understanding of the limits of science. The idea that science can produce definitive answers to all questions, given adequate resources, is no longer tenable. Science now tells us that some things are fundamentally unknowable. The full impact of this change has not yet begun to be felt in the formulation of public policy.

The past few years have witnessed the modern equivalent of a witch hunt in which products are driven from the marketplace without trial on the basis of suspicion. At the same time, we have allowed continued production of products that are clearly damaging the environment but for which ironclad evidence is lacking. Our challenge is to reform the policy process and inform the populace so that some degree of rational behavior can exist in the presence of vast uncertainty.

What Can Science Tell Us About Environmental Risk?

The science of environmental analysis has made remarkable strides in telling us what is present in our environment. Analytical chemistry and

Table 6.1 Trace Analysis Units

	1 part per million	*1 part per billion*	*1 part per trillion*
Time	1 minute/2 years	1 second/32 years	0.06 second/2000 years
Salted potato chips	1 pinch salt/ 20 pounds	1 pinch salt/ 10 tons	1 pinch salt/ 10,000 tons
Martinis	1 drop vermouth/ 80 fifths gin	1 drop vermouth/ 500 barrels gin	1 drop vermouth/ football field–size pool of gin 43 feet deep

related disciplines have invented and refined the tools of modern analysis over the past three decades. The limits of detection of many substances have improved from the part per million (ppm) level to the part per trillion (ppt) level, an improvement of a factor of a million or more. These concentrations are so low that it is hard to conceive of them at all. Table 6.1 gives some examples of the relative magnitudes of these terms.

It is now possible to find almost anything anywhere. In fact, for some substances it is possible to detect single atoms. We know that there are detectable quantities of *enriched* uranium in the spring water from which some brands of beer are brewed. We can find aluminum everywhere. Alar and its carcinogenic breakdown products can now be detected in apple products. The same carcinogen, unsymmetrical dimethylhydrazine (UDMH), is naturally present in mushrooms. This analysis was not possible when Alar was licensed or for at least 20 years thereafter. It took millions of dollars of Uniroyal money to develop the methods that ultimately forced them to drop the product. The range of potential carcinogens in our fatty tissue is large indeed and includes substances such as DDT that have been banned from the marketplace for more than a decade. On the global scale, we have remote sensing of pollutants that warn us of a diminishing ozone layer. We have a great body of scientific information on global problems such as acid rain and the accelerating greenhouse effect. We are inundated with information about the presence of hazards. Progress in analysis has not stopped, and there is still much that needs to be done. However, we must now decide what to do with all this information.

What Are the Limitations of Science?

Developments on the second front, our understanding of the nature of the risk posed to people and the environment by all of these chemical substances, have not been nearly as promising. Which of the 60,000 synthetic chemicals in the environment are dangerous? How do they hurt people? Do they cause damage to the ecology? It is these areas in which science is lacking.

Gathering risk analysis data regarding human exposure is not morally or ethically possible except in cases of accidents. Therefore, we must work by extrapolation from model systems. Popular sentiment against the use of laboratory animals is rising. While there is no doubt that animal studies are important, there is also considerable doubt about the accuracy of extrapolations to humans. In the case of threats to the global ecology, we have only computer models and small-scale experiments to guide us, and they often produce conflicting results.

Let us look at one specific area of uncertainty and see how it produces problems for public policy. It has been said that the only chemicals that will not ultimately be proved to be carcinogens within the current legal definitions are those that are very toxic. These will kill the test animals before they become old enough to develop cancer. Current environmental regulations require that new chemicals be tested at a sufficiently high level to show an effect. In fact, the testing must be done at the maximum tolerated dose (MTD), which is the highest level that does not kill the animals.

There are three types of testing currently in use. One, the Ames test, uses a specific strain of bacteria to test chemicals rapidly for possible carcinogenic activity. Those that do not yield a positive Ames test have very low potential as carcinogens. The Ames test and other screening tests seem to be good at raising questions and identifying substances of concern. They cannot provide answers about human risk.

The second type of testing uses animals. Animal tests are a problem for several reasons. First, these tests are very expensive. The cost of screening a single compound can exceed one million dollars. It can cost over $10 million to conduct the tests required to register a new agricultural chemical for use. Costs for a pharmaceutical product are as much as ten times higher. The second problem is the lack of good extrapolations between species. Just because a substance is a carcinogen in one animal, one cannot predict what will happen in a second species. An example of this problem is found in the toxicity data for dioxin. The term dioxin is often used to describe a family of related chemical compounds. The most toxic of these is TCDD (2,3,7,8-tetrachlorodibenzo-p-dioxin). Dioxins are the product of combustion that occurs in fireplaces, trash incinerators, and trash-to-energy plants. This family of chemicals is often used to stir up the NIMBY reaction to the latter two installations. Dioxins were also the chemicals present in Agent Orange, which was thought to have caused so much damage to American servicemen in Vietnam.

The toxicity data for dioxin in a number of species are given in Table 6.2. Imagine that you are an executive at Dow Chemical. You have

Table 6.2 Dioxin Toxicity by Species

Animal	*Lethal Dose (microgram/kg body weight)*[a]
Guinea pig	1
Rat (male)	22
Rat (female)	45
Monkey	<70
Rabbit	115
Mouse	114
Dog	>300
Bullfrog	>500
Hamster	5000

[a]The lethal dose (LD_{50}) for a chemical is the amount that kills half of a group of test animals.

Source: Poland and Knutson. Adapted, with permission, from the *Annual Review of Pharmacology and Toxicology*, Vol. 22, copyright 1982 by Annual Reviews Inc.

commissioned a study on male rats, and when you see the results you realize that you have a problem chemical on your hands. So you order a second study on a very different animal, bullfrogs, because they are cheap to study. These additional results would produce nothing but confusion because bullfrogs can obviously tolerate large doses of dioxin. So you order studies on guinea pigs and hamsters. Now you are really in trouble. Clearly, the toxicity varies by several orders of magnitude. Different species presumably detoxify this compound in different ways. What is the human sensitivity? We still don't know. What do you tell your boss about the hazards of your products 2,4,5-T and Agent Orange? What do you tell the Congress and the EPA about the hazards of dioxin?

There are a number of assumptions involved in animal studies. It is necessary to find animals that respond to a specific toxin in the same manner as people (this assumes all people respond in the same way, which is questionable). The response to a given toxin can vary dramatically, as shown in Table 6.2 for dioxin. Choosing an appropriate test animal is, therefore, quite difficult. In fact, until one knows the nature and mechanism of the toxic reaction it is impossible to make an informed choice of a test animal.

Primates, because they are the closest to humans in biochemical function, would be the best models. However, they are expensive to study. Studies on primates also raise the greatest public outrage because of their similarity to humans. Rats and mice are cheap and do not produce the same negative public response to their use as test animals. However, in some ways they are not similar to people in their biochemistry.

Another problem is that it is difficult to assess risk at very low levels of exposure. Suppose we wish to determine the amount of aflatoxin B, a possible carcinogen found in peanuts, that will cause an increase of one part in one million in mortality in rats. We could raise one group of a million rats on a control diet and another group on a diet that contained a known amount of aflatoxin. After a fixed time, we could determine how many rats died from each group. If we assume that 50 rats died from the control group and 60 from the aflatoxin group, this result would imply that the dose of aflatoxin we administered caused about 10 excess deaths per million rats. We still would need to estimate the amount that would cause one part per million more deaths. Would that be one tenth as much aflatoxin as was used in our experiment? That assumes a linear relationship between dose and hazard.

In the example above, two million rats would be required to conduct the study, and this would be enormously expensive. We would need to buy and feed the rats, possibly for several years. Even then there would be a statistical uncertainty in the result; if one repeated the study again, it is doubtful that the results would be exactly the same.

The only practical solution to the problem of measuring the effects of a toxin is to increase the dose, which increases the response level so that it can be detected in smaller groups of test animals. We could, for example, increase the dose of aflatoxin by a factor of 10,000. If we assume a linear model for the response of rats to the toxin, we would only need to have 100 rats in each group.

This is a practical solution, but it raises serious questions about the validity of our extrapolation to low doses. There are any number of possible relationships between the dose of a hazardous substance and the response of the organism. Some general possibilities are shown in Figures 6.1 and 6.2. Figure 6.1 shows a linear-dose response relationship. As one can see, risk doubles when dose doubles. The bottom line shows a threshold below which there are no effects. The middle line has a zero risk only at zero dose. The top line shows the results that are obtained when background risks cannot be eliminated from a study. Figure 6.2 is a simple non-linear model for dose-response relationships. Quadratic, cubic, and exponential functions increase much more rapidly at high concentrations but actually lie below the linear model at low concentrations. In reviewing these figures, it is clear that one must know not only the shape of the curve but also where one is on the curve before an accurate estimate can be made.

Most studies of this type begin with a linear model of response. There is no alternative that makes any more sense than the linear model as a starting point. Exposure to ionizing radiation appears to follow a linear model, and the use of this model for chemical exposure is based on this fact. Also, the linear model is simple to understand and to work with.

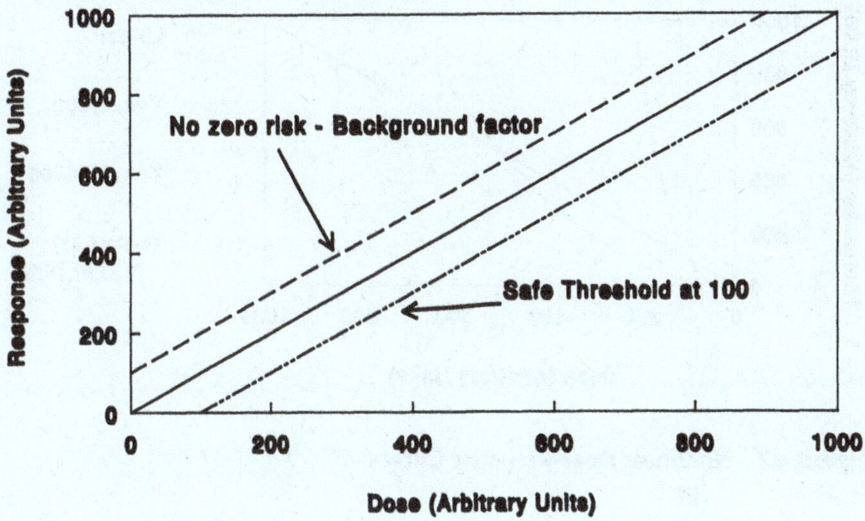

Figure 6.1 Linear Dose-Response

The one major study of chemical dose response curves that has been carried out (Staffa and Mehlman 1980) used 24,000 mice over a four-year period and cost over $12 million. During the study, 90 tons of mouse food were used, and the total staff for the study was about 150. Even this investigation, which dealt with only a single chemical, 2-acetylaminofluorene, could not establish whether there was some threshold level below which the chemical was no longer a carcinogen.

Recent analysis of the 2-acetylaminofluorene experiment (Cohen and Ellwein 1990) has provided strong evidence that the current use of rodents for the evaluation of the carcinogenic potential of chemicals is fundamentally flawed. This study and the work of Ames and Gold (1990) support the idea that any chemical that causes cells to grow at a greater rate than normal will produce excess cancers.

The dose-response curves that result from the effects of chemicals on cell proliferation are very non-linear. Typically, this mechanism produces a threshold below which no cell damage occurs and no cancers are produced. Above this threshold, the rate of cancers increases very rapidly with a pronounced upward curvature that ends at the toxic level where the animal dies. Because almost any chemical at nearly toxic doses results in cell death and rapid replacement, one should expect most chemicals to cause cancer under the conditions now required for these tests. In fact, data for the chemicals that have been thoroughly tested in both rats and mice show that 212 out of 350 synthetic chemicals and 37 out of 77 natural chemicals in food are carcinogens (Ames and Gold 1990).

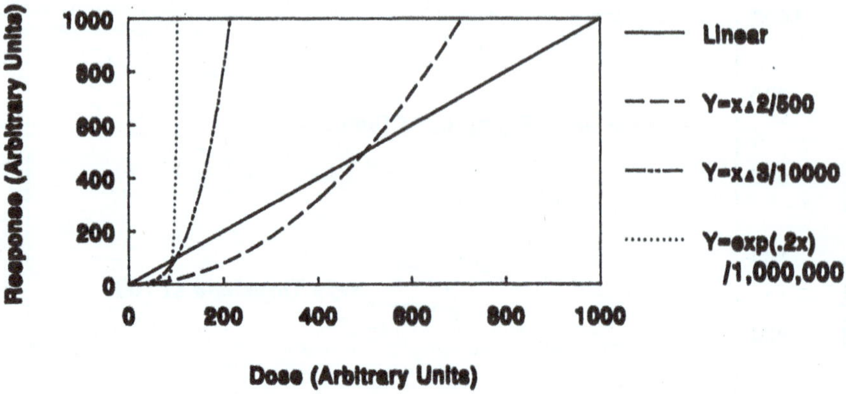

Figure 6.2 Nonlinear Dose-Response Curves

If one were to assume that these animal tests predicted human cancer risk, one would have to conclude that people are at much greater risk from eating the natural chemicals in carrots, apples, grapes, orange juice, and a host of other foods than from the traces of synthetic chemicals in the food supply. The known natural carcinogens in our diet are over 20,000 times more prevalent than synthetic carcinogens.

The implication of this finding is that current risk analysis and the public policy that grows from it must be radically altered. New methods of classification and testing will have to be developed or else we will be forced to abandon most uses of chemicals.

In some cases, the mechanism by which an organism deals with a toxin at high doses may be different from that which occurs at low doses. This can also lead to discontinuous dose-response curves. One interesting example is formaldehyde. Formaldehyde air pollution has been a cause of public concern because of the use of formaldehyde in wood products and fabrics. Dose-response data for the production of nasal cancer in rats who inhaled different concentrations of formaldehyde are given in Figure 6.3. The cancer rates are based on a 24-month study with 160 rats in each test group. No tumors were observed at 0 or 2 ppm. It appears that there is a threshold level below which no risk of cancer is observed (Starr 1990). The explanation for this is based on chemical reactions that occur in the mucous layer in the nasal passages. Formaldehyde is very reactive, and when it acts with the mucous, it forms a harmless substance, formate. It is only when the concentration of formaldehyde reaches some threshold level that it can penetrate the mucus and attack the tissue below.

The formaldehyde example demonstrates the need for an enormous amount of information before a true understanding of a toxic hazard is possible. The risk must be identified. In this case, the risk is the promotion of nasal cancers by formaldehyde. The mechanism of action or detoxification must be determined, and the presence of this mechanism in various species must be measured. Only then can we begin to use science to make good public policy. However, the interim between identification of the hazard and a full understanding of risk can be very long indeed.

The full story on formaldehyde is not yet known. It appears that formaldehyde bound to dust particles has a greater effect than that present as a vapor in the air (Gammage 1990). While it is back to the laboratory for the scientist, what can be done in the meantime to protect people from the *potential* hazards they face from formaldehyde in the environment? Can we wait until we are certain that we know what we are doing? Public pressure regarding environmental issues does not allow that option. Morally and ethically we cannot allow it either. The only choice is to act without full information.

In conclusion, it will be necessary to radically change the way we conduct animal studies. The results of these tests at high levels can provide little guidance about the true risk of a chemical. Tests at low to moderate levels may allow us to deal with the most serious risks first. However, everyone must come to realize that, even in ideal

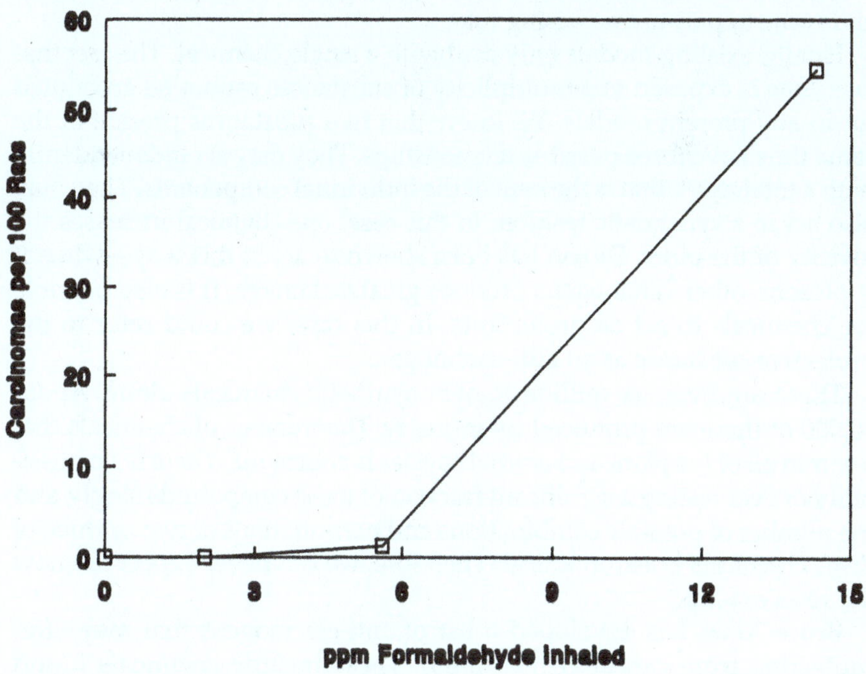

Figure 6.3 Formaldehyde Dose-Response
Source: Starr, 1990.

cases, the risk assessments derived from animal studies are estimates, not answers.

A third method of assessing risk that has the potential to produce both more rapid and more reliable results is currently under development. This technique uses actual human DNA and proteins to assess the damage caused by toxins. Strong carcinogens appear to bind to or alter DNA, and this can be measured quite well. These compounds are called genotoxic, and the degree of genotoxicity appears to correlate well with the ability to induce cancers. These compounds also appear to best fit the linear dose-response model with no safe threshold.

One finding of the early application of this method was that the effects of chemicals can vary considerably from one individual to another and from one organ to another. This variation raises the possibility of needing large numbers of different human DNA samples to obtain a valid result. Care must be taken in selecting these samples so that they adequately represent the diversity of the population. These studies also ignore the fact that living systems can behave in ways that are more than the sum of their parts.

In spite of the potential limitations of DNA and protein-based tests, they probably represent the best tool for risk assessment. As we learn more about the mechanisms of carcinogenesis, these tests will be refined and come to play an increasing role.

Finally, existing models only deal with a single chemical. The fact that everyone is exposed to a multiplicity of substances cannot be accounted for in any present models. We know that two substances present at the same time have three possible relationships. They may act independently with a total result that is the sum of the individual components. They may also act in a synergistic fashion. In this case, one chemical increases the toxicity of the other. Dioxin has been shown to act in this way—when it is present, other carcinogens produce greater damage. It is also possible for chemicals to act as protectants. In this case, we could refer to the protective substance as an anti-carcinogen.

There are over six million known synthetic chemicals alone. About 60,000 of these are produced by industry. The number of chemicals that occur in all of the plant and animal species is enormous. There is no possibility of ever testing a significant fraction of these compounds singly, and the number of possible combinations and permutations of two or three of these chemicals is astronomical. Therefore, we can never expect to know the risks exactly.

Bruce Ames has developed a list of anti-carcinogens that may offer protection from cancer (Ames 1983). These include chemicals found naturally in foods such as carrots, broccoli, eggs, and garlic. On the other hand, he has also compiled a list of natural food carcinogens. These include the compound in mushrooms mentioned previously as well as chemicals in beans, fruits, vegetables, and grains. It is clear that the combination of multiple factors can have either a positive or negative impact on the total organism. It is not clear how the total cancer risk is affected by diet or what fraction of the risk is due to synthetic chemicals.

If we want to be consistent about our concern for our safety, we should probably forget the risk from pesticides and other synthetic chemicals and worry more about the hazards of our voluntary activities. Smoking causes more deaths than all of the other forms of air pollution combined. This fact is now well established for everyone not employed by the tobacco companies. The current estimates of the risk from naturally occurring radon gas in the outside air is that it appears to be at least 10 times more hazardous than EPA or FDA allowable food additives levels. Houses with indoor radon levels equal to the current federal standard produce a risk of lung cancer of between 1 in 100 and 6 in 100 (Byrd 1990). These are large risks. They are not the result of human action, but some measures could be taken to reduce them.

Only when we have dealt with the large risks should we turn our attention to the chemicals in foods. This is not intended to justify the continued use of genotoxic chemicals. It does mean that we should not waste resources on tests that do not help us determine the mechanisms of the actions of chemicals. We have reached the point where we can no longer trust simplistic animal testing, but we have not fully developed the alternatives.

How Does This Uncertainty Affect Public Policy?

The Delany Clause, which bans any known carcinogen from the food supply, and the mentality that produced it are major parts of the problem we now face. In the 1950s and 1960s, it seemed possible to guarantee absolute safety in the food supply. At that time we could not detect all of the hazardous chemicals that were present, and we had not identified many of the compounds that we fear today as dangerous. In our ignorance we believed an absolute standard could be imposed.

The present state of knowledge in the regulation of potential hazards makes such an absolute standard impossible. The analytical chemists will measure the minute quantities of a host of potential hazards, natural and synthetic, and others will quantify the threat that these substances pose to the limits of their models. Then we must decide as a society just what level of risk we are willing to accept. Only then will there be a way out of the regulatory thicket.

The current efforts to rationalize the regulation of pesticides and other crop protection chemicals is an excellent example of the problem. The EPA estimates that the major portion of the risk of cancer caused by chemical residues on foods is posed by chemicals that were registered for use long before the current level of technology existed. At the time these chemicals were approved for use, they passed all possible tests that were available and required, but many of them could not meet the more rigorous standards of today. Unfortunately, we know that many new crop protection chemicals cannot pass the absolute test of the Delany Clause. If we cannot bend this rule, we will be left without any chemicals to support the production of food for the population. Although it is possible to make a good case for a move to a less chemically based agriculture, this is not something that can be done quickly without major disruption of the food supply. During the late 1980s, the EPA found itself faced with the fact that the way to reduce the risk of chemical residues in food was to bend or break the Delany Clause.

The Role of Science in the Decision Process

Science still has a role to play in the regulatory process, but we ought to be aware of what science can and cannot do.

One thing that science can do is to communicate risks to the general public to allow decisions based on the information that is available. It is equally important to communicate the uncertainties in the assessment of the risks. Scientists are often not as good at this, and the language of uncertainties can be confusing even to scientists. The education of the public about risk must include attention to both the risks of modern technology and the background risks present in the natural environment. The latter is a particularly important and underappreciated aspect of risk communication. The public should understand that decisions must be based on moral and ethical rather than purely scientific grounds. The public needs a context for moral and ethical judgments.

The one thing that science cannot provide is absolute answers about risks. Many people look to science to provide these answers, and this demonstrates a basic misunderstanding of the nature of science. Science deals with proving which things are false. It can never prove something is absolutely true because it cannot prove that it has tested all possible options. The greater the number of possible explanations for a phenomena that are proved to be false, the higher the probability that one of the remaining explanations is in fact true. But if no one can conceive of a possibility, scientists cannot test it. In a sense, we are limited by our metaphors and the context in which we work. These define the range of questions that we can formulate.

The flurry of hypothetical causes for Legionnaires' disease a few years ago is a useful case study. Everyone seemed to have a theory, but many of these were tested and found to be wrong. No one was sure until the causative bacteria was identified. Along the way, the disease was attributed to a variety of causes—bacteria, viruses, and pollutants such as nickel carbonyl.

In the same way that science cannot prove anything is true, it is not possible to prove anything is 100 percent safe. Science can only determine that something is probably safe. Substances that are really dangerous are fairly easy to identify. Cyanide is definitely toxic—we have substantial proof of that. Alcohol is probably unsafe. The evidence for this is compelling, yet we continue to use it. But out at the fringes of hazard, the questions are more difficult. Saccharine and cyclamate are two examples of substances with which science has great difficulty. Ultimately, we are limited by our imaginations as to possible hazards. Only recognized hazards can be tested.

The current debate over Alzheimer's disease is an interesting example of our lack of imagination. The cause of this terrible and expensive disease is not known. Because aluminum is found in high concentrations in the brain lesions, it could be that our use of aluminum in foods, water purification, pharmaceuticals, and cosmetics is to blame. Fluoridation of water, that old villain of the extreme political right, may be responsible because aluminum concentrations in cooking increase dramatically in the presence of fluoride. Or the cause may be a virus. Even common proteins such as glutamate in MSG may be responsible. Finally there may be a portion of the population that is genetically predisposed to Alzheimer's disease and is more sensitive to some of the above chemicals than the population at large. Certainly, no one thought of these possibilities during the regulatory approval of the products now in question. Even if it were possible to evaluate these products using humans as test subjects, how could scientists assure that every possible genetic group was included in the test?

Many practices that now produce outraged calls for regulation were begun because people could not imagine the types of hazards involved. Because they couldn't imagine the hazards, there was no reason to test for them. As new knowledge about biochemistry emerges, new fears emerge to be evaluated. Furthermore, as the tools of science improve, it becomes possible to test new ideas in ways that were beyond the power of earlier scientists. This can produce some unsettling results, both for the scientists and for the public.

In conclusion, if anything is clear from all of this, it is that the regulatory process must exist in a constant state of flux. Decisions must be made on the basis of insufficient information in order to protect people. When better information becomes available, there must be ways to adapt earlier decisions without recriminations. We should expect mistakes to be made in any regulatory process that uses scientific information as its basis.

What Values Result from Our Efforts at Scientific Risk Assessment?

Scientists are often heard to complain about the public's lack of appreciation of risk assessment. The public also appears to dismiss much of what scientists say about risk. Why is this? There is a basic difference in the way risk is viewed by the public in general and by those who attempt to quantify risk. A number of psychological studies have demonstrated this fact (Slovic 1987). It can be easily argued that if people behaved in a manner consistent with scientific risk assessment, there would be fewer untimely deaths. People would avoid risky behavior in favor of safer alternatives.

Probably the best example of the disparity between the way risk is viewed by scientists and by the public is the continued smoking of cigarettes by over one-quarter of the adult population. The risk of getting cancer in this country is now about 0.33, which means that one-third of the population will eventually get this dread disease. The major single source of cancer is cigarettes. This fact has been demonstrated as clearly as science can ever demonstrate that something is true. We have epidemiological data, the results of animal studies, and a mechanistic understanding of the way in which smoking causes cancer. Still, we find smokers demanding the removal of Alar from apples. The cancer risk of Alar in apples is equal to or less than the risk these people willingly expose their children to when they smoke in their children's presence. The EPA bans most uses of benzene, which causes leukemia, but a smoker exposes me to unpermissible levels of benzene from sidestream smoke. No one has yet been able to prevent this known risk. The recent decision by the EPA to list sidestream smoke as a carcinogen is a step in the right direction. Their estimate of three thousand cancer deaths per year from this source makes this a significant threat.

Scientific risk analysis places equal value on all risks. Dying is dying, and how one gets there is not part of the risk model. People do not think that way. Instant, spectacular death is much feared. Air travel is possibly the means of transportation most feared by the general public. Yet an analysis of the real risk shows that this is one of the safest modes of transportation available. Most people accept the risk of car travel, and the majority of the population does not even bother to use the seat belt. This may be because we are familiar with the risks, and the accidents that do occur are less spectacular. We also feel we have more control because we are at the wheel of our own car. The same is probably the case with smoking. Smokers take a small risk every time they light up, but there is a chance of survival too. Each little action can be ignored, and the long-term consequences of the action are lost in the pleasure of the moment.

The consequences of smoking and of driving a car are also not as visible as the news coverage of a plane crash. What if the bodies of the 1,000 smokers who die every day from lung cancer and heart disease were shown on the nightly news each day? Or if the 100 or more daily auto fatalities were all shown on the national news. Would there be some impact? Would there even be time for any other programming?

A number of studies have shown that our greatest fears are catastrophic risks and hidden risks. This accounts for the public fear of air travel and pesticides, respectively. We accept familiar and obvious risks, like driving a car, with little thought.

There are two courses of action for the scientific community in the face of this seeming irrationality. We can try to educate people to behave as they "should," based on our science. This means helping people understand the values on which scientific risk assessment is based. Alternatively, scientists can try to adapt risk analysis to include the alternative set of values expressed by people in the way they live. This means finding a way to insinuate the biases of the population into the analysis. Probably the right answer is to do some of both. The chemical industry is finally awakening to the need to educate people about risk. This is an important step in the right direction.

When regulators dismiss the values of the population in making public policy, they are doomed to failure. Therefore, risk analysis must adapt to the public values. At the same time, policy makers and scientists should work to help the public understand the scientists' values in assessing risk. Ultimately, the standards of acceptable risk will not be consistent for all risks.

Risks to the Global Ecology

Most of the discussion thus far has focused on risks that affect only a small population. The use of a pesticide that causes 100 additional cancer deaths each year has little or no effect on the rest of the population. However, risks to the global ecology have the potential to affect everyone. The nature and magnitude of the consequences vary with where one happens to live rather than with what one eats or drinks. These risks do share something in common with those previously discussed in that they appear distant to most people because they take a long time to develop.

On the global scale, some risks appear to be very large indeed. Possibly the best example of the need to formulate policy based on ethical and moral values rather than on scientific knowledge is the accelerating greenhouse effect. Science has predicted the *possibility* of an environmental disaster for over one hundred years. Svante Arrhenius made the first published calculations of the effect of doubling the carbon dioxide in the atmosphere and arrived at a change of about 4 degrees Celsius in the earth's temperature. This falls within the limits of uncertainty of the best modern estimates. The current estimates are more convincing because they have tested a much larger number of alternatives. A considerable research effort is being mounted to refine these estimates and model the results. However, it is not reasonable to expect science to achieve complete confidence in its estimates until it is too late to make major changes. By the time we build confidence in any scientific prediction, we will only make policy to deal with the costs of the problem. Prevention or amelioration will no longer be an option.

A public policy to deal with greenhouse gases raises very difficult moral and ethical questions. What value can we place on the quality of our lives in the industrial West relative to the quality of life in the Third World? Is the life of an American more valuable than that of an Ethiopian? Should Americans eat beef fed with grain, while others starve for lack of grain? Should the rain forests of Central America be burned to make space to graze cattle? Should poor farmers in Central America be prevented from raising cattle in order to raise their standard of living? How much growth is sustainable? How can the resources of the earth be shared equitably?

These are difficult questions, and they make many of us uncomfortable. It would be easier to accept the answers if we knew with certainty what they were. But if we do not face the questions, that very action represents our answer to the questions.

In some cases, where the scientific answers are already clear, the scientific values conflict with other values. An example comes from the effort to remove chlorofluorocarbons from use. We know with high confidence that these chemicals are destroying the ozone layer, which protects living things from damage by ultraviolet light from the sun. The ozone layer is already decreasing significantly, and there is preliminary evidence that susceptible species are already dying as a result. The nations of the West have agreed to phase out the use of the chemicals by the year 2000 or sooner.

The conflict in values arose in 1989 when China was beginning a major program to build refrigerators. Two technologies were available, one using chlorofluorocarbons and one using helium. The new helium technology is as harmless and environmentally sound as we can expect refrigeration technology to be. China had applied to the United States for a loan to allow it to purchase the helium-based technology. Unfortunately, the Chinese government chose to repress the demonstrations in Tiananmen Square at that time. The values of democracy and freedom demanded that the United States not provide financial assistance to the Chinese government, and the refrigerator program was begun using chlorofluorocarbons. The environmental value of eliminating the Chinese use of chlorofluorocarbons for the twenty- to thirty-year lifetime of the refrigerators was lost.

This is one example of the conflict in values that will arise as nations attempt to deal with the global ecological problems that face us. Clearly, even good scientific evidence can be offset by strong political or moral values.

Conclusion

Science can now tell us, with a high level of confidence, exactly which chemicals in what amounts are present in our environment. The ability of science to determine the exact effects of these chemicals on our health, especially on the long-term production of cancer and other diseases, is very limited. In the case of global problems such as the destruction of the ozone layer and the accelerating greenhouse effect, large uncertainties can be expected in the scientific predictions until it is too late for effective action. Furthermore, it is not reasonable to expect that this situation will change significantly. Although methods will continue to be refined, it is unreasonable to expect further improvements in scientific methodologies to make significant contributions to the process of making policy. In fact, the better our scientific understanding becomes, the more complicated the policy choices may be.

As a society, we have to face difficult moral and ethical choices about which risks we choose to accept. Unfortunately, there has been a wide gap between scientific risk assessment and public perception. The scientific community has exacerbated this by insisting that the public accept its values regarding risk. The underlying problem is that a scientific assessment of risks often ignores the fears and desires of the general public. Although a death is a death from the objective scientific perspective, from the human point of view there are many ways to die and they are not all perceived as equally frightening.

Science has a major role to play in understanding the environmental impact of human activities. It is crucial that quantifiable risks are identified and irrational fears laid to rest with correct information. Only then can decisions be based on rational choices. It is also important that scientists listen to the fears of those without technical training, for in their imaginations we may find new risks that we have missed. Society must come to realize that complete freedom from risk can never be achieved.

The promise of Delany Clause absolutes are a fiction and must be replaced by a reasonable acceptance of risk. Finally, action must often be taken on limited information; it must be possible to implement consistent public policies without a demand for scientific certainty. Industrial lobbies and environmental action groups must ultimately learn that they have a common interest in the intelligent use of technology and the reasoned acceptance of some risk.

References

Ames, Bruce. 1983. "Dietary Carcinogens and Anticarcinogens." *Science* 221: 1256.

Ames, Bruce N., and Gold, Lois S. 1990. "Too Many Rodent Carcinogens: Mitogenesis Increases Mutagenesis." *Science* 248: 970.

Byrd, Daniel M. 1990. "Risk Assessment and Risk Management." New Jersey Clean Air Council public hearing, April 17, 1989, as reported in American Chemical Society, Division of Analytical Chemistry Newsletter, Fall.

Cohen, Samuel M., and Ellwein, Leon B. 1990 "Cell Proliferation in Carcinogenesis." *Science* 249: 1007.

Gammage, Richard B. 1990. "Exposure to Formaldehyde in Indoor Air." *Risk Analysis* 10: 77.

Poland, Alan, and Knutson, Joyce C. 1982. "2,3,7,8-Tetrachlorodibenzo-p-Dioxin and Related Halogenated Aromatic Hydrocarbons: Examination of the Mechanism of Toxicity." *Annual Review of Pharmacology and Toxicology* 22: 517.

Shodell, Michael. "Risky Business" *Science* 85:46.

Slovic, Paul. 1987. "Perception of Risk." *Science* 236: 280.

Staffa, J. A., and Mehlman, M. A. 1980. "Innovations in Cancer Risk Assessment" *J. Environ. Path. Toxicol.* 3:1.

Starr, Thomas B. 1990 "Quantitative Risk Estimation for Formaldehyde." *Risk Analysis* 10: 85.

PART THREE

Environmental Values and the NIMBY Syndrome

7

Partisan Politics, Economic Growth, and the Roots of NIMBY: The Case of Montpellier, France

William C. Gunderson

As attested to by conflicts between states producing high-sulfur coal and states producing low-sulfur coal, conflicts between solid-waste-producing and solid-waste-consuming communities, and friction between hydrocarbon and renewable energy enthusiasts, the tension between economic growth and environmental pollution manifests itself in many settings. The problem has traditionally been technically defined in terms of benefits and costs (e.g., the jobs saved by allowing a "dirty" factory to produce its goods versus the long-term public health consequences of that production), making it a candidate for resolution through an appropriate mix of financial incentives. But the essential conflict might also be understood in political terms.

Environmental problems provide political entrepreneurs the opportunity to sling mud at opponents. (The benefits of negative campaigning—e.g., the "Boston Harbor" strategy of the Bush campaign in 1988—have been demonstrated again and again.) Thus, as important as the trade-off between economic growth and environmental pollution is, to focus solely on that trade-off is to presume an extremely rational political process (i.e., one based on technical considerations as opposed to fear). If would-be elected officials take partisan advantage of a nagging problem (exploit NIMBY fears), then fairly sophisticated benefit/costs analyses cannot be expected to play an important role in decisional contexts.

Furthermore, from the perspective of those who wish to undermine an incumbent, economic growth itself becomes a political matter. Rather

To aid in the use of citations in the text references and minimize space, I will use an abbreviation system. For example, *DL* means Démocratie Locale and *LG* means *La Gazette*. A complete list of the abbreviations leads the reference section of this chapter on page 126.

than being viewed as a benefit that can be used to convince people to tolerate the costs of environmental damage, economic growth will be judged by some citizens as a cost in the first place.

This attitude yields an interesting twist to our benefit/cost understanding of these matters and creates a cost/cost, not a benefit/cost, situation. That is, economic growth, rather than being perceived as a benefit that mitigates the costs of environmental damage, is viewed as a cost that is responsible for another cost (environmental despoliation). Indeed, as suggested by recent events in Montpellier, a major city located in the south of France, economic growth as a public policy focus can create tensions that make a dispute over the siting of a landfill worse than it would have been in the absence of that growth.

Local Politics in France

Before exploring the possibility that economic growth and environmental pollution can best be understood in cost/cost rather than in benefit/cost terms, some background on local politics in France is in order (Schmidt 1990).

Since the 1870s, mayors have had considerable clout in the French system, in part because of the home rule powers granted to them by the central government and in part because they are able to gain stature in a political party by becoming national figures. (They can hold seats in the National Assembly and be a mayor at the same time.) This status is further enhanced by the party slate mode of election. Mayoral candidates are listed first on their party slate. If the slate wins, the candidate listed first becomes mayor. Having won, the mayor has a built-in majority to support his projects (Schmidt 1988).

The impact of having this majority can be seen at city council meetings where legislators often roam the room and chat with colleagues or read newspapers while business is being transacted. To help his party members "tune-in" to the proceedings, the mayor raises his hand indicating that it is time to vote.

The power of a mayor is also reflected by the informal means by which motions are brought to the floor. When the author visited Prades, a small town in the mountains near the Spanish border, he noted, while reading city council minutes, that important changes in land use policy were presented orally by the mayor and changed significantly from one meeting to the next.

For example, at one point the interests of downtown merchants were protected through a resolution limiting commercial development along the new highway by-passing the city center. (The by-pass allowed traffic from the coast easier access to ski areas in the Pyrenees.) Within a few weeks, this policy was changed to allow additional commercial development near the by-pass. All of this was done at the behest of the mayor (Prades 1988).

This power which has existed for several decades, is being enhanced by some recent trends. Whenever one talks about the smallest political unit in France, the commune (we will refer to them as "towns"), someone usually says, "There are 35,000 of them and they must be consolidated!" This desire to reduce the number of towns and, thereby, realize economies of scale is encouraged by the European Community, which believes that member states should have local governmental units of comparable size.

To respond to these pressures, Paris is providing greater flexibility to large cities by encouraging them to consolidate services with surrounding towns and thus create districts. In Montpellier, where economic growth and its environmental impact have become issues, a district provides transportation and fire protection as well as solid waste disposal services. The mayor of Montpellier is also the president of the district.

Because of a mayor's ability to move beyond the territorial confines of his city and gain access to additional development sites and tax resources through his control of the district council, he becomes a threat to small-town mayors whose jurisdictions he can influence through the powers of the district. If the mayor's political party is different from the political parties of some of the mayors whose towns belong to the district, conflict may be exacerbated.

Departmental and regional councils, as well as districts, provide new options and sources of conflict for mayors (DL 1985). The department is the territorial division of the French state used by Napoleon to consolidate the power of the center vis-à-vis the periphery at the end of the eighteenth century. The state was represented by a prefect, a quasi-military figure who maintained law and order and coordinated the state's activities in the periphery.

Over the past few decades, but particularly since the decentralization reforms of the early 1980s, the executive powers of the prefect have been transferred to the president of the department's legislative branch—the departmental council. (As is the case with a mayor, the president of a departmental council heads a party list of would-be departmental legislators.) This role provides a new power base and source of funds for an aggressive mayor. The difficulty is that a mayor may or may not control the departmental council's president. Therefore, this newly enhanced role may cause him as much grief as opportunity.

Similarly, France's new level of government—the region—may help extend a mayor's policy and budgetary options or provide a launching pad for his opposition. Régions were not created until 1972, and their councils were popularly elected for the first time in 1986.[1] They help coordinate economic development by creating a plan every five years and signing a contract with the state that commits the latter to provide funds for education, tourism, infrastructure—projects identified in the five-year plan (*Région* 1989). As is the case with the department, the head of the party slate that wins control of a regional council becomes the regional president and its chief executive officer.

Montpellier and the Thôt

An excellent example of the possibilities created by and the tensions emanating from district, department and regional councils is illustrated by the conflict between Georges Frêche, a member of the Socialist party, and some of the mayors who work with a right-of-center opposition to Frêche and sit on the district council.

On February 24, 1989, Michel Vaillat, the mayor of Lattes, and several members of his executive committee chained themselves to the fence surrounding the Thôt, a landfill created by the city of Montpellier in 1967 (*ML* 2/25/89; *MVV* 2/90, 3). What had begun as a cooperative endeavor between Lattes and its "big brother" Montpellier, located just a few miles to the north, had evolved into a highly charged political issue and classic NIMBY problem.

Although the landfill was sanctioned by the previous mayor of Montpellier (François Delmas) and his counterpart in Lattes (*QC* 4/90, 3), the current occupants of those roles (Georges Frêche, mayor of Montpellier since 1977, and Mr. Vaillat) had fought over the landfill long before the February confrontation. It is important to note that Georges Frêche and Michel Vaillat are members of different political parties (PS, Socialist Party, and UDF, Union for French Democracy, respectively). But to fully appreciate how highly politicized this conflict has become, it is important to understand the broader context within which it has emerged.

Georges Frêche's Strategy of Economic Growth

An aggressive program of economic development has created considerable tension between Montpellier and the surrounding towns. Under the leadership of Georges Frêche, Montpellier has adopted an economic growth strategy that local leaders hope will transform the city and surrounding area into one of the six to eight French "Eurocities" of the early twenty-first century (Di Guisto 3/21/90; Daume 1990, 661). A sign that the rhetoric of economic growth ("Montpellier, the supergifted") is not empty can be seen in the city's population growth from 80,000 to 207,000 in about twenty-five years.

At the center of Georges Frêche's development strategy are a series of technopoles. The technopoles take advantage of existing resources, such as the research capacity of an academic staff needed to service over 40,000 university students, one of the oldest medical schools in Europe (founded in the fourteenth century), the largest private employer in Montpellier (IBM), centers for agricultural research, and recreational and cultural assets. These resources are combined with sizable government subsidies and private-sector interest and energy to encourage product development. The new products, in turn, enable Montpellier to compete in the emerging European and global markets.

In short, the landfill controversy and its partisan and NIMBY dimensions must be placed in a larger context where a young mayor (Mr. Frêche was in his thirties when first elected) has grabbed a quiet southern convalescent center by the throat and is dragging it into the twenty-first century. His penchant for big projects and grandiose names—Antigone, Port Marianne, Millennaire, the Corum—in addition to his leadership style (to say he is aggressive and undiplomatic would be understatements) exacerbate tensions that are a consequence of rapid economic growth.

Regarding grand public works, the technopoles represent millions of dollars of investment. A multi-purpose conference and entertainment center, the Corum, built to strengthen the tourism technopole, cost nearly 150 million dollars.[2] But this is the tip of the iceberg. As early as his first mayoral campaign, Mr. Frêche had signaled his interest in "big projects."

When Georges Frêche ran for mayor in 1977 as a member of the University of Montpellier's law faculty, he promised to "cross the Lez," a river located just a few blocks east of the center of town. Development from the center toward the east had been blocked by a one-hundred-acre military camp established by Louis XIII early in the seventeenth century. (The king created a garrison city after the local citizenry revolted against the crown [Margé 3/9/90].) Georges Frêche claimed that only he, not the status-quo leadership, was capable of ridding Montpellier of this monument to the past and of launching the projects necessary to guarantee the city an important role in the new Europe.

Upon election, Frêche secured the one hundred acres, hired a well-known architect from Barcelona, Ricardo Bofill, and created Antigone, a monument to neo-classical splendor containing government offices, public housing, shops and public spaces stretching from city hall and an urban mall at the eastern edge of downtown to the "new east," the *hôtel de région* (the headquarters for Languedoc-Roussillon's regional government) located just across the Lez (*MVV* 1989, 13).

Soon the technopole centered around IBM, located to the north of the regional headquarters, and a yet-to-be-built complex that will include a marina, a new city hall, and a golf course/residential development, located to the south, will one-up Antigone. It will extend urban development farther east and will go beyond linking downtown Montpellier to the river Lez—it will open the city to the sea four miles away! To accomplish this, the Lez must be dredged so that pleasure boats can navigate from the Mediterranean to the proposed marina (Puissance 1988).

Reaction to Frêche's Leadership and Fiscal Style

If the grandiose dimensions of the mayor's development efforts are not enough to exacerbate built-in political tensions, his leadership style is. He manifested his aggressiveness at the end of a tough meeting on the budget by bragging that "my technique is like a fragmentation bomb; I explode all over the place." He has demonstrated his lack of diplomacy by telling adversaries in public that they "do not understand a thing" (*LG* 5/5–11/90, 3; *ML* 2/1/90, x3).

Aggressiveness and a lack of diplomacy can be augmented by denying old-fashioned political resources to factions that do not cooperate. When the Greens (the political party concerned primarily with environmental protection) told Frêche that they refused to help him implement his solid waste program, he told them that they would not "take part in" his administration. (It was clear that "take part in" meant getting special budgetary consideration and other favors [*LI* 2/22/90].)

Frêche's willingness to let people know who is in charge is further illustrated by his presidency of the SERM (Societé d'Equipement de la région Montpellieriane), the joint public/private venture (one of the largest in France) that oversees virtually all of the city's development projects. By placing the responsibility for implementing his growth strategy in a semi-private agency, he can play his cards close to his vest and frustrate active oversight.[3]

The penultimate example of the mayor's willingness to take charge, often undiplomatically, can be found in the Canourgue incident. When city-center residents objected to the mayor's plan to turn a small park into a parking lot and threatened to copy the mayor of Lattes' gesture at the Thôt by chaining themselves to the park's trees, Mr. Frêche had a half-dozen city employees sneak into the park at four o'clock on a Sunday morning and preempt the protest by dispatching the trees with buzz saws (*LG* 10/13–19/89, 5–7).

Even though the mayor's "go-go" development strategy and personal style has attracted as many supporters as detractors (Mr. Frêche won re-election in March 1989 by getting over 50 percent of the vote in the primary and a year later, Montpellier won a "most dynamic city" prize given to one of the 212 French urban centers having more than 100,000 inhabitants [*ML* 3/29/90, x9]), the mounting costs of making Montpellier a "Eurocity" has many citizens worried.

In May 1989, after April's electoral campaign, the few members of the district council who regularly criticize Georges Frêche complained that he delayed consideration of the budget for several weeks for political reasons. (It is usually approved well before April.) When combined with the previous year's budget, it represented an 86 percent increase. If one goes back to 1980, one discovers that the total of city and district budgets equalled $20 million. The total for 1990 was $146 million, or a 750 percent increase (*LM* 4/11–25/90; *ML* 5/15/89).

Despite these significant budget increases, major projects (e.g., waste water and solid waste treatment, mass transit), which other cities the size of Montpellier consider important, had been placed on the back burner. To begin to compensate for some of these "oversights," Montpellier committed about $28 million early in 1990 to begin upgrading a waste water treatment plant to accommodate the metropolitan area with a population projected to grow from 260,000 to 340,000 by the turn of the century. Unfortunately, some observers fear that this effort will only patch an antiquated system rather than create something that will adequately serve the metropolitan region in the years ahead. In short, many ask whether long-term investment in basic services are being sacrificed for the mayor's grandiose plans (*ML* 3/29/90, x4).

Although most of these critical observations come from opposition politicians, others who have no partisan axes to grind have expressed concern for Montpellier's rate of growth and the environmental and fiscal implications of that growth. Some of the CEOs who have been attracted to Montpellier from Paris over the past several years by the quality of life are concerned that the rapid pace of growth and its costs (noise, traffic, solid waste disposal problems) are undermining the ability of the city to attract new firms. On the tax front, some labor-intensive firms are concerned that even the traditional activities of the district, such as the metropolitan mass transit system, are forcing considerable tax increases. For example, taxes on a labor-intensive firm employing 400 to 500 employees increased from $120,000 in 1989 to $180,000 in 1990 (Médevielle 1990).

Conflict Between City Hall and Regional and Departmental Councils

These tax increases may be more than the Frêche administration anticipated given their hope that the city would get help from the regional and departmental councils. In placing the *hôtel de région* at the opposite end of Antigone from city hall, the mayor signaled his expectation that he would control the former as well as the latter and thereby have a ready source of funds for his projects. Unfortunately for the mayor, a right-of-center coalition under the leadership of Jacques Blanc won the 1986 Regional

Council elections. The new majority demonstrated its antipathy to the mayor by reversing a previous decision to give Montpellier $10 million for the Corum (*LG* 3/30/90–4/5/90, 10–11).

Antagonisms between the mayor and the president of the regional council have reached the point where the city has refused to host any of the sports during the 1993 Mediterranean Games, a major initiative of the region. The mayor claims that Jacques Blanc has favored other towns with his Mediterranean Games largess and has left Montpellier to fend for itself regarding sports infrastructure (*LG*3/2–8/90, 5).

As if his poor relations with the regional council were not enough, Frêche's *nouveau arrivé* origins have alienated those in the Socialist party who control the departmental council. (The departmental council has the deepest pocket of all the local governments. For fiscal 1990, the city planned to spend $340 million, the district $112 million, the department $620 million and the region $387 million [*LG* 2/16–22/90, 8].) The current president of the departmental council, Gérard Saumade, is a close ally of former National Assembly deputy and president of the Hérault Federation (the Socialist party's departmental organization), Gilbert Sénès (*ML* 2/16–17/90, 2). Mr. Sénès's past presidency of the departmental fair board, an organization that annually helps celebrate the wine culture, is illustrative of one of the party's traditional bases of support—the wine growers (Lepouriel 1990).

By way of contrast, Georges Frêche is an academic non-native of Montpellier and devotee of the newly emergent pro-growth/liberal faction of the Socialist party. He has neither the Sénès/Saumade ties nor the policy orientation to endear him to Hérault's traditional party base. He was born in 1938 in Toulouse, received a diploma from HEC, the most prestigious business school in France, and a doctorate of law degree before joining the law and economics faculty of the University of Montpellier. He was elected to the National Assembly in 1973 and became mayor of Montpellier in 1977 (*ML* 3/22/90, w1).

In addition to Frêche's academic and managerial background, his decision to support Rocard over Mitterand at the 1978 Metz party congress and the debacle at the 1990 Socialist party congress at Rennes where Frêche and Saumade were supporters of competing factions did nothing to mitigate intra-party tensions in Hérault (*LM* 4/11–25/90, 10; *LM* 3/1–15/90, 6–7). Thus, political opposition from both the regional and departmental councils has forced the mayor to seek other fiscal means, including tax increases, to finance his projects.

What we have then is far from a "simple" landfill problem. We have a dispute over solid waste disposal that emerges in the context of ambitious development plans put into motion by a young, take-charge mayor. Moreover, his leadership style is not calculated to calm the fears of those concerned about the environmental and fiscal impacts of an effort to make Montpellier one of the urban centers of France.

The Politics of Landfill Siting

Having established the broader context within which a classic partisan/NIMBY problem can play itself out, let us turn to the landfill controversy itself. Although some observers believe that the existing one-hundred-acre site is the best available and should be used until 2050 (*LG* 3/16–22/90, 6–8), others, particularly the inhabitants of Lattes, want the old landfill shut down.

Opponents of the present site claim that as early as 1967 the prefect placed a cap on the landfill at eight feet, a limit that was surpassed in the early 1980s. (It is now over thirty feet high [Bentegeac 1989; Berthezene 1990; *Nature* 1990, 319].) During the 1977 election, Georges Frêche asserted that the landfill was a problem created by his predecessors. When he ran for re-election in 1983, he supported an incinerator as the best option. Given growing concerns about the air pollution caused by this method, the mayor promised to find a better alternative if elected to a third term in the spring of 1989 (*LM* 3/15–28/90, 7).

Frêche made good on his promise by announcing in May 1989 that he had decided to open a new landfill in Montarnaud, a small town located about eight miles west-northwest of Montpellier. Through the District of Montpellier Council, the legislative body traditionally concerned with solid waste disposal, he offered to purchase about 1,400 acres where a collection, composting and burial operation would be located.

As pleased as the mayor of Lattes and his associates were by this announcement, the mayor of Montarnaud and the mayors of the towns surrounding the proposed site, responding in part to NIMBY fears, moved to block the project. Opposition to the plan focused on poor site selection and loss of potential development.

Regarding site selection, the mayors of the four affected towns[4] brought in consultants from throughout France as well as from Spain and Belgium, to argue that the topography of the proposed site is not adequately impermeable and a major source of drinking water will be threatened if the project moves forward. Supporters of the site agreed that some fissures exist, but argued that proper engineering could isolate it from the surrounding area (*LG* 3/16–22/90, 6–9).

In addition to the impermeability problem, opponents believe that the site's elevation makes it susceptible to considerable damage by thunderstorms and rapid run-off. To protect the site from this danger, millions of francs will be needed to build drainage ditches that can channel water safely away from the site (*ML* 2/15/90).

To highlight the loss of development potential, the four mayors presented a major project that would use the land for an Olympic-level training and sports medicine complex (*ML* 1/15/90). The proposed center (projected cost of $120 million) would provide facilities for several sports and the supporting diagnostic and physical therapy capabilities to attract athletes who regularly compete internationally.

That two other towns in the department had already submitted similar proposals did not bode well for the Montarnaud project. One proposal came from Lattes and the legislative district of Willy Diméglio, the right-of-center opponent of Georges Frêche in the 1989 mayoral race and close ally of Jacques Blanc, president of the regional council. The second proposal estimates costs of only $10 million and has been endorsed by Gérard Saumade, president of the departmental council (*ML* 2/10/90, 2). Of course, the proposal's purpose may have been largely "political" in that it attempted to highlight, particularly for the citizens of the four towns surrounding the site, the opportunity costs that they must bear if the new landfill is built.

Given the strong negative reaction to the Montarnaud landfill site, one might ask whether the mayor and his staff anticipated the NIMBY reaction to their site selection decision. It is fairly clear that they did. Throughout the 1980s over a half-dozen reports were presented by various government and private agencies concerning options to the Thôt. Without exception, the options were presented with major reservations, indicating that each option carried significant increases in costs and considerable uncertainty regarding potential environmental damage. The mayor accused the authors of these studies of ducking under an umbrella to protect themselves from the flack they would get if they offered definitive advice.

Despite all of the uncertainty, Georges Frêche gave in to the tremendous political pressures and promised to do something about the problem if re-elected for a third term. (Part of the pressure was self-imposed, given his concerns about the presence of the old landfill in the golden triangle, an area south of Montpellier extending to the sea that includes several of his cherished projects.) He took a leap of faith.

Although a prefectoral report issued in late 1989 recommended a large regional incinerator with co-generation at Sète/Fontignan located ten miles west of Montpellier (Bentegeac 1989), the mayor dropped that option given the electoral success of the Greens and their adamant opposition to incineration. (The mayor predicted that within a few years the Greens would be able to get 15 percent of the vote in local elections, while the Communist party would be able to get only 5 percent (*LG* 2/2–8/90].) Because another option, the gas-generating process know locally as

Valorga, had not yet proved itself viable in pilot tests,[5] he settled on "tri-selective" with composting and burial.

The success of this $44 million tri-selective project depends upon the willingness of citizens to separate their garbage and place it in different colored containers. As part of the educational process, the city council approved on March 23, 1989, a collection and separation center in the Paillade, a neighborhood in Montpellier. Through this process, glass, paper, plastic and metals will be separated for recycling at three locations (the home, the neighborhood center and the district center in Montarnaud). This part of the process will eliminate 30 percent of the waste stream.

The second part of the process will involve composting the organic residues and selling them as fertilizer, an activity that will remove another 30 percent of the waste. The remaining 40 percent will be compacted and buried in the forty-acre site selected for that purpose (DM 1990).

Even though many people supported the new solid waste disposal method, the debate over the old and the new sites continued. At the same time, a special commission convened by the prefect considered whether or not to grant the Declaration of a Public Utility requested by the district on June 6, 1989 (DM 1989, 29–31).[6] On April 20, 1990, the commissioners did an extraordinary thing—they said neither yes nor no in their report.

The opponents of the new siting immediately claimed victory, suggesting that the commission reflected the doubts expressed by the inhabitants of the four towns and all of the experts who criticized the new site. Although the opponents knew that the battle was far from over, they believed that they had won a major symbolic victory from the commission.

Equally surprising was Georges Frêche's reaction. He chose Earth Day, April 22, 1990, to announce a referendum, the first in France focusing on the environment. It would not give the voters an opportunity to choose from several options but rather would allow them to respond yes or no to two propositions: (1) Do you support "tri-selective?" and (2) Do you support the selection of the Montarnaud site for composting, compacting and burial? The mayor argued that there had been so much heat generated by the controversy (each pressure group claimed to have the public interest at heart) that it was important to get a grass roots reading on the matter.

To those who charged that the mayor was ducking his responsibility, he responded that he should not be so accused because he had had the courage to make a choice after several years when no other public official had stepped forward to recommend a specific site and a specific technology. He added that he was growing tired of the demagogy surrounding the issue and had decided to get the politicians out of the picture for the moment.

Others claim that the mayor's decision to hold a referendum on the landfill controversy was a wise political move, designed to get him off the hook, given NIMBY pressures and the maneuvering of his opposition. If the voters endorsed the new site, he could forge ahead with popular support. If not, he could stay at the Thôt, declaring that that was his inclination all along. The latter option would at least save the taxpayers a considerable sum of money.

The mayor of Lattes and his associates were very concerned that it was precisely the cost factor that would encourage the voters to say no to the proposed site, effectively keeping the Thôt opened. Therefore, they were adamantly opposed to the referendum, declaring that the matter must be taken to the Conseil d'Etat for final resolution (LCC 1990; Berthezene 1990).

Conclusion

Montpellier's landfill controversy involves conflicts common to large metropolitan areas where waste must be disposed of but where virtually no public officials are willing to sanction a particular technology or site given the risks, both political and technical.

On the political side, it is certain that the opposition will use such a problem to embarrass the incumbent. Furthermore, leaders of small communities, even though they may be allied with a big city mayor, find it necessary to stand up to the "encroaching" power in order to maintain their credibility among their constituents. (Both Charles Maneiro and Alfred Conesa, members of the Socialist party and two of the four mayors whose towns are closest to the proposed site, had been endorsed by Georges Frêche during the spring 1989 mayoral elections ([*ML* 2/6/89].) Adding to the political problems are the technical difficulties associated with predicting how fiscally and environmentally costly a given disposal method will become.

Of course, these political and technical problems are made much worse because the landfill dispute has evolved in the context of the aggressive (and undiplomatic) development strategy of the mayor of the dominant city. As complex as the technical matters are, they are rendered secondary given the existing tensions between Georges Frêche and the mayors of several small towns located in Montpellier's shadow.

An interesting dimension of this case concerns the idea that all that is needed to overcome the NIMBY problems associated with landfill siting is to provide an appropriate mix of financial incentives. That assumes that such a problem is subject to calm, rational benefit/cost analysis. However, what is happening in Montpellier suggests that given the struggle for power (both within and between political parties), financial incentives are irrelevant to the key actors.

For them, the notion that political processes provide opportunities to solve community problems (e.g., find the least costly means to get rid of garbage without too much environmental damage) is not uppermost in their minds. Rather, the fact that Montpellier's mayor is being forced to select a new landfill site and solid waste disposal technology gives them an opportunity to manipulate NIMBY fears and overturn his 1989 mandate.

As appropriate as it might be to understand economic growth and environmental pollution in benefit/cost terms, it is also clear that conflicts generated by an aggressive, pro-development mayor may allow his political opposition to cast both growth and pollution in negative terms. Thus, in order to launch an image-destroying electoral strategy, what is normally understood as a benefit/cost problem is translated into a problem in which one piles costs on top of costs.

Notes

1. France has twenty-two regions and ninety-six departments. Montpellier is the capital of a region, Languedoc-Roussillon, consisting of five departments located at the northwestern corner of the Mediterranean. All but one department border the Mediterranean starting west of Marseilles and extending west and south toward the Spanish border. In addition to being a regional capital, Montpellier is the capital of Hérault, one of the departments located in the eastern part of the Region. Languedoc-Roussillon has a popluation of two million and is half again as large as Maryland (three times the size of Connecticut). Hérault has a population of just under 800,000 and is two-fifths the size of Maryland.

2. For conversion purposes, we will assume five francs to the dollar.

3. An illustration from the United States of the use of quasi-governmental agencies to "get things done" by means that are at least partially out of the public eye may be seen in the career of Robert Moses. See Robert A. Caro, *The Power Broker* (New York: Random House, 1974).

4. Charles Maneiro, Montarnaud; Alfred Conesa, Murviel; Claude Seignier, Saint Georges d'Orques; and Paul Lassalvy, Saint Paul.

5. A pilot project was being run in Amiens but was suspended until new loans could be negotiated. The costs of the conversion from solid waste to gas had risen to over eighty dollars per ton (that compares with the sixteen dollars per ton it costs to place solid waste at the Thôt), and the gas produced was not good enough to use, even though the city had already invested $27 million (*ML* 3/23/90, m3).

6. The district council made four decisions on the proposed site—accepted the study report, #1288; agreed to purchase land for the site, #1289; requested the Declaration of a Public Utility from the prefect, #1290; and authorized a study of the landfill site.

References

To reduce the space required to make references in the text to the citations that follow, the following conventions will be used:

DL	Démocratie Locale
DM	District de Montpellier
LG	*La Gazette*
LCC	Lattes City Council
LM	*Le Montpellierain*
LI	*L'Indépendant*
ML	*Midi Libre*
MVV	*Montpellier votre ville*
QC	*Que Choisir*
Région	*Région Languedoc-Roussillon*

Bentegeaç, Yves-Jean. 1989. "Propositions en vue du traitement des Déchets Urbains de l'Est de l'Hérault." January 10.

Berthezene, G. First Vice-President of Lattes City Council. Interview, April 26, 1990.

Daume, Daphné (ed.). 1990. *Britannica Book of the Year*. Chicago: Encyclopedia Britannica, Inc.

Démocratie Locale. 1985. "1982–85: trois ans de décentralisation." *Bulletin De La Direction Générale Des Collectivités Locales, Ministère De L'Intérieur Et De La Décentralisation*. #38, April.

Di Guisto, Elvio. President of Euromedicine, one of Montpellier's five technopoles. Interview, March 21, 1990.

District de Montpellier. 1989. "Séance Ordinaire." June 6. pp. 29–31.

District de Montpellier. 1990. "Inauguration de la Déchetterie de la Paillade." January 27.

La Gazette. 1989 and 1990.

L'Indépendant. February 22, 1990.

Lattes City Council. 1990. Special meeting. May 5.

Lepouriel, Guy. Retired army officer and close observer of local politics. Interview, May 5, 1990.

Margé, M. Department of City Planning, City of Montpellier. Interview, March 9, 1990.

Medevielle, Marc. *Midi Libre* journalist. Interview, April 6, 1990.

Midi Libre. 1989 and 1990

Montpellier votre ville. 1989. "Vue par la presse internationale." p. 13.

Montpellier votre ville. #128, February 1990, p. 3.

Le Montpellierain. 1990.

Nature. April 1990, p. 319.

Prades. *Registre des délibérations du Conseil Municipal*. March 17, 1988, p. 17; June 27, 1988, p. 24.

Puissance: *Le Journal D'Information Du District De Montpellier*. #39, April 1988.

Que Choisir. #260, April 1990, p. 3

Région Languedoc-Roussillon. 1989. "Contrat De Plan Etat—Région Languedoc-Roussillon, 1989–1993."

Schmidt, Vivien A. 1988. "Unblocking Society by Decree: The Impact of Governmental Decentralization in France." Paper prepared for delivery for the Research Committee on the Comparative Study of Local Government and Politics at the International Political Science Association International Meetings. Washington, D.C. August 31–September 1.

———.1990. *Democratizing France: The Political and Administrative History of Decentralization.* Cambridge: Cambridge University Press.

8

Environmental Values, the Economic Ethos, and NIMBY: The Rhode Island Case

Talbot Page

This chapter addresses three questions: (1) What is the landfill siting problem in the state of Rhode Island? (2) To what extent does a case study of the landfill siting problem support Kelman's (chapter 2, above) view of separate domains for environmental values and public policy analysis? (3) How does one teach a nitty-gritty case study on landfills and develop from it some more general ideas for public policy analysis? Addressing the first question helps get at the second. And how we address the third shapes our understanding of the first and second. The three questions form a package.

The case study described here was undertaken by one of my undergraduate classes of environmental studies students at Brown University, most of whom would identify themselves as having "environmental values." I had the students use an economic approach to public policy analysis. Thus the class project itself provides a case study on the separate domains of environmental and economic perspectives. What follows here is a report on that class project, and as such it is interpretative and anecdotal. As part of the class project, the students wrote a report of their own (Albert et al. 1990). Compared with this chapter, the student report is more focused on problem solving—developing policy recommendations that might be used in the regulatory process.

According to Kelman, environmental values are roughly those values held by John Muir. In environmental values wilderness is valued highly; environmental values are "less competitive, less oriented toward money,

I wish to thank my students in ES 192 and especially the teaching assistant Michael Muller. They have helped a great deal in thinking through the course on which this chapter is based. I hope that I have not misstated their views.

less instrumental," compared with economic values (Kelman, this volume). Pinchot, who valued and managed forests for their use value, provides a paradigmatic example of someone holding and promoting economic values. According to Kelman, economic values are held by many policy analysts, especially those trained in economics. These differing value orientations lie in separate domains. A person may value things from the perspectives of the two domains, sequentially, without making a blend of "trade-offs" between the two domains. A tension arises when policy analysts use economic approaches to evaluate environmental regulation. Policy analysis itself becomes an attack on environmental values.

The case study offers some support for Kelman's view that there are two domains and a "tension" between them. At the same time, the problem of siting a landfill is hardly the same as that of protecting a wilderness area, and the relation between the two domains in the case study appears to differ from the one suggested by Kelman. In my effort to explain the interaction of the two domains, I will address the three questions posed at the beginning of this chapter.

Core Ideas of Policy Analysis

There are many possible approaches to policy analysis, and more particularly to environmental regulation. The one I followed was based on three ideas: (1) the conceptual separation of positive and normative questions (I first separated normative and positive analysis and then asked how the two related to each other in the case study at hand), (2) an informal version of game theory (an example of this approach can be found in McCubbins and Page [1987]), and (3) a Bayesian approach to uncertainty and learning (using informal value-of-information). The approach, with these three core ideas, does not require assuming that individuals have (exogenously given) complete preference orderings or that social valuations are defined in terms of summing willingnesses to pay or the "measuring rod of money." Thus it differs from the cost-benefit analysis discussed by Gillroy in this volume or exemplified by Baxter in *People or Penguins* (1974). For example, if a student believed that recycling had a "special" value as a "symbolic expression of caring for the environment," a value that should not be treated as a "personal preference" evaluated according to a compensating variation, the approach to policy analysis taken in the course does not directly attack this view. But as we shall see, there may be some indirect tension.

The NIMBY[1] issue posed a dilemma for my students, who were well aware of the NIMBY syndrome before taking the class. They tended to think that opposing a landfill site on the grounds that it was "in my backyard" was selfish and hence inconsistent with environmental values, which were viewed as more communitarian. However, they also realized that local opposition to landfills makes landfills harder to site and more expensive, makes waste processing more expensive, and thereby promotes source reduction and recycling. Because source reduction and recycling are high on the list of environmental values, there was an ambivalence about NIMBY. The students were also sympathetic to the NIMBY cries from the poorer communities, which were in danger of having the site forced upon them. Thus there was no clear-cut environmentalist evaluation of NIMBY preferences and consequentially no environmentalist analysis of the siting problem, because dealing with NIMBY preferences was at the heart of the siting problem.

I started out on the policy approach by telling the students on the first day of class that we would focus on two questions: (1) How can one predict and explain where the state of Rhode Island would locate the next landfill? and (2) Where should the site be located? In this way, I started out with a separation of positive and normative questions (and I was not challenged on this separation during the rest of the term). I then passed out outline maps of Rhode Island and said, "We can save a lot of time if you just mark an X where you think the state will locate the next landfill and an O where you think it should."

There was a moment of silence as the students stared at the blank maps and wondered if this course was for real. Clearly, the task as posed was not doable. I waited for a few awkward minutes to let this sink in and then asked what we would need to know in order to answer the "is" and "ought" questions as they came to be called. We quickly obtained a shopping list of things we would need to learn in order to answer those questions.

The students wanted to know the location of wetlands, aquifers and surface waters and their current qualities and uses; soil conditions; the locations of roads and population centers; the state of the art in landfill technology, containment success and monitoring; and the possibilities of remediation. In short, the students wanted to learn the same sorts of things as the technical experts who were helping the Solid Waste Management Corporation (SWMC) to identify sites.

The students also wanted to learn about legislation, regulations, and the roles of the principal institutional actors. They wanted to know what bothered people most about landfills—traffic, odors, groundwater contamination, inadequacies in monitoring and remediation, or what? Thus they wanted to know the same sorts of things as the policy people who were also advising the SWMC and the political people who were shaping the regulatory process. Going further, the students wanted to know the history of the siting problem and what was happening in other states. They had heard that eminent domain was an issue important to the siting problem, and they wanted to understand what that issue was about.

This list, which was surprisingly easy to elicit, provided a research agenda we were able to use and to a large extent follow for the rest of the course. From this list, we defined tasks that we assigned to groups of students, or "task groups," who would study these matters and report back to the class as a whole. The tasks included creating a flow chart of the key players and processes, and researching the following: the experiences of other states, the history of the landfill siting issue as reported in the last thirty years of the *Providence Journal*, eminent domain, exclusionary criteria and geology, computer mapping systems, the market approach and the compensation issue, the experiences of the Narragansett Native Americans (who were tentatively negotiating about a site), and normative issues. To encourage the students to organize their thoughts before they did the research of the task groups, I asked them in the first week to write three- to five-page papers responding to the "is" and "ought" questions.

The "is" and "ought" questions, the research agenda and the group tasks set up the course as an exercise in means-ends rationality, where the things to be learned on the shopping list were important because they instrumentally aided in addressing the "is" and "ought" questions I posed at the beginning.

This sort of scholarly means-ends rationality is of course not limited to policy analysis. One can find such rationality in many fields, for example, in trying to find what it takes to prove a mathematical theorem or in searching for the date of a manuscript to establish literary priority. Thus I followed an approach to public policy analysis that overlaps with other disciplines. Clearly, the approach differs from those used in English or mathematics by its treatment of normative and positive questions and by its emphasis on game theory and Bayesian assessment of uncertainty. But the point is that the three core ideas—separation, game theory, and Bayesian assessment—are (or can be) strongly oriented toward a means-ends rationality. Characterizing the domain of policy analysis in this broad way, by the three core ideas, helps establish a partial overlap with other fields and may reduce some of the tension Kelman talks about.

To provide a focus for the course as a whole, I told the class that by the end of the term we would, as a class project, write a regulatory paper, which would analyze the landfill siting problem and make recommendations. If successful, it would be readable and useful to regulators, scientists, politicians and others involved in siting decisions. This project became the student report.

Informal Game Theory

I asked each student to read approximately forty-six pages of Rhode Island statutory law and then construct a decision flow diagram of the siting process. This diagram, which looked like a map or wiring diagram, represented the institutional rules of the game (it took many hours for students to construct these maps). In constructing the maps, we tried to identify the key players, which included: Solid Waste Management Corporation (SWMC), which would pick two or three new sites and manage them; the State Planning Council, which would certify that the Corporation's plan fit the statewide comprehensive plan; the Siting Board, which approves the Corporation's choices; the Department of Environmental Management (DEM), which ensures that the environment is protected and licenses the facility; the governor, who may have to use eminent domain to secure the sites; the state legislature, which wrote the law and established the process in the first place; the cities and towns, which may or may not be able to exercise exclusive zoning or other local veto power; and the experts and consultants, many of whom are hired by SWMC, who do the studies and develop the "decision methodology."

While identifying the players, we tried to identify as well the possible actions they could take ("strategy sets"), their motivations ("utilities"), and the most important uncertainties. We talked only a little about formal game theory, but we used its conceptual framework. We analyzed the one-shot prisoner's dilemma, enough to see how individual rationality can lead to social pathology. We noted that the pathology was from the prisoners' point of view. From the point of view of the detective who modified the game form by adding the possible "trumped-up charges," the prisoner's dilemma was a clever system of incentives, designed to lead to good social outcomes (the prisoner's confessing).

We tried to figure out the consequences that would result from particular collections of players' actions ("the game form"). (The game form was hinted at by the students' process diagrams of the institutional rules). The effort to uncover the simultaneous causality of the game form gave their diagrams a more interpretive feel than the diagrams of the statutes and agencies we came across during the case study.

I encouraged (somewhat strategically) the students to be as complete and detailed and comprehensive as they could be in their process diagrams. This emphasis seemed natural to the students, who wanted to be "realistic" and "practical," in contrast to economic modeling approaches, which they sometimes called "abstract," "theoretical," "misleadingly simplified," and "impractical."

The goal of realism meant close reading of the statutes (which was fine by me), but even with close reading there were ambiguities and things we did not know how to interpret. For example, one clause in the Siting Act required the SWMC to maximize resource recovery and recycling. Did this provide legal access to a town trying to block a site on the grounds that the state could not locate a landfill until it had first maximized reuse and recycling? Or was there no real veto threat in this language because the SWMC could argue that it was a national leader in recycling programs? As another example, the Siting Act eliminated two Blackstone Rivers from consideration as sites, but there is only one Blackstone River in Rhode Island. Was this simply a mistake? How did this and other exclusionary criteria get written into the legislation anyway? Compared with the tipping fees for industrial waste and compared with other states, the fees for municipal solid wastes were low ($13/ton). The floor for compensation to the host community was set at $1/ton or an aggregate of $7,500, whichever was greater. But how much greater could the compensation go? Compensation could be a part of the bargaining process after the site was chosen, but could it be a part of the site selection process itself?

To address these and other questions of interpretation, we brought in four outside speakers. The first was Sue Kiernan from the groundwater office, DEM. She provided historical background. She told us that forty years ago nearly every town had its own dump, often located on what were called "useless swamps" or in other out-of-the-way places. During the 1970s and early 1980s, many of these dumps were closed by DEM to protect groundwater. Now there are only nine operating landfills left in the state, and of these only a few important ones. The most important one, Central Landfill at Johnston, has the largest volume (4,000 tons/day) in New England. By satellite it is the most visible structure in Rhode Island and is one of the highest elevations in the state.

Ms. Kiernan told us that Rhode Island is one of the few states where the need for new landfill sites is not prompted by the existing facilities becoming physically full. The Johnston landfill still has thirty years of capacity, but after a period of confrontation politics during which the mayor of Johnston (aRusso) slowed down the haulers for "safety inspections," the governor (DiPrete) agreed to close Johnston as a landfill by 1994.

Ms. Kiernan spoke of the many contentious hearings concerning Johnston and other siting decisions. She said that although groundwater protection was an important concern, people often spoke of more immediate concerns, such as truck traffic (600 large trucks come to Johnston each day), odor, litter, and property values.

Our second speaker, Dante Ionata, assistant director of planning, SWMC, helped us interpret the legislation. The original agreement was to present the legislature with a package (developed by the Environmental Quality Commission) that the legislature could vote up or down without amendment. But this agreement came apart, and amendments were added at the last minute. The exclusionary criteria were added at the end, a great deal of the exclusionary list came from a legislator from the northwest part of the state. Many of the rivers listed in the exclusionary criteria are in that part of the state, as is the Blackstone River, which, in the legislator's haste to make sure this river was listed, was listed twice. We also found out that a river that does not exist in Rhode Island was also listed.

Ionata viewed compensation as a fairly cut-and-dried, minor part of the act, which comes into play after the site is chosen. The requirement for maximizing resource recovery and recycling was added at the end, and he did not know what legal opportunities it presented (he doubted it would make much difference because there is similar language elsewhere and the state has a strong program in recycling).

And then, just as Ionata was finishing up, we asked him what his decision diagram of the process looked like. In fifteen seconds he startled the class. He said that SWMC must make the site selection and provide the justification for its decision, but then this decision must be agreed upon by four agencies along a sequential path. He drew a picture on the board of a straight line with four possible veto points on it. This drawing was of course much simpler than our own "realistic" maps of the process. It said to us "one lead player and four possible vetoes." Was this stylized version the "real game" and more realistic than our own detailed maps? We asked him the obvious question: "What was the chance that the delegated process would not lead to site selection [as shown on our maps] but would lead to impasse and perhaps back to Johnston?" "Very high," replied Ionata, "but if the choice came by default back to Johnston, the compensation paid to Johnston would be very much higher than now."

Ionata's last minute in class had a big impact. Maybe the search for complete detail and full complexity was not the best approach. Maybe the detail cluttered what was "really" going on rather than revealing it. I pointed out that nothing is so useless as a 1:1 physical scale map. A map becomes useful when it emphasizes the most important things and abstracts from the less important.

From previous classes, I had the impression that environmental studies students were suspicious of simplified economic models and valued more realistic (and complicated) descriptions. But now, prompted by Ionata's comments, we discussed how we needed simple, stylized stories that could help us interpret the "real world," much as a road map helps us interpret the outside world. We needed not one but several stylized stories, so that we could assess competing interpretations or models. Ionata provided us with one. We needed to identify others. With a list of possible models, we could think about what evidence we might look for in order to discriminate among them.

We read "A Theory of Congressional Delegation" (McCubbins and Page 1987) in order to generate ideas about how delegation games might arise and how they might lead to the veto story Ionata briefly sketched. The idea of the paper is that legislatures delegate to executive, administrative and regulatory agencies for several reasons, including complexity, resource and time constraints, the need for "political daylight," and so on. But whatever the reasons for delegation, the form of delegation is influenced in predictable ways by two factors.

The two factors are uncertainty and conflict (among the constituents and the legislators themselves). If a particular bill exhibits a more than average level of either factor (or of both factors), the issue is perceived by the legislators as a "hot potato." The hotter the potato, the greater the incentive to pass it on to an agency; the bigger and more definitely specified the mandate, the more extensively are the tools of action listed (for the agency); the bigger the break between the principal and the agent, the more "political daylight" between the two.

But with this increase in delegated power comes a tendency in the opposite direction. The more the uncertainty and conflict, that is, the hotter the issue, the greater is the risk that the legislature is creating a "loose cannon." To prevent the agency from becoming a loose cannon, the legislature builds safeguards into the enabling act itself. Consistent with the issue (e.g. landfill siting) being a "hot potato," these safeguards are preferably ones that can be effected at arm's length (from the point of view of the legislature)—safeguards that can be implemented "automatically" or by others.

The preferred safeguards are procedural in nature. They prescribe hearings and procedures that, if not followed, invite judicial review by others. They allow for sign-offs and certifications by other agencies. With more conflict and uncertainty, the lead agency is given a bigger mandate; but then others are given access, through procedures, and this access can amount to vetoes and requirements for modification of the lead agency's proposed actions.

In a sense, when there is more conflict and uncertainty at the legislative level, more is given with one hand and more taken away with the other. But this is not quite the right way of putting it, because what is given and what is taken away are different.

After reading McCubbins and Page's paper, it seemed to us that although there might be less uncertainty associated with landfills than with toxic wastes or some other issues, there was a great deal of conflict among the towns as potential host communities. For comparison, the situation can be contrasted with the air toxics legislation. This legislation was not considered controversial when it was created. It was expected that DEM would operate pretty much in the shadow of the U.S. Environmental Protection Agency and that not much would happen that was not happening already. DEM was given a broad and somewhat vague mandate, with relatively few resources to achieve it and relatively few procedural safeguards to keep DEM out of trouble.[2] Compared with the less controversial (and perceived less conflictual) air toxics program, the landfill siting issue seemed to involve more conflict, as evidenced by the resulting ambitious delegation of power and the confining procedures specified by the legislature. Did the landfill siting process boil down to the "hot potato" described conceptually by McCubbins and Page and in practical terms by Ionata? If so, was the siting process born to lose? Failure was a fairly likely prediction from this model.

We could not satisfactorily address these questions until we had considered other interpretative versions of the institutional process (game form). The papers assigned in the first week of class suggested some alternative explanations. In answer to the "is" question (where will the state locate the next landfill site?), by far the most common student response was that the state would locate the site near a politically weak and economically poor town. The idea was that the legislature acted sufficiently as a unitary body so that it delegated enough power and could send a clear signal for SWMC to "tag the weakest player" (and for the governor to make it stick through eminent domain if need be). This second stylized story also suggested predictions: the siting process would be successful in that a site would be chosen and a list of towns would be determined (because they were politically weak).

As an alternative to this story of unitary action, we considered a spatial voting model that explicitly takes into account the divisive nature of NIMBY preferences reflected from voters to legislators. Imagine that for each town, the worst possible site is right next to the town itself. As the possible site gets farther away from the town, its desirability increases, until the site is ideally far away from the town. The ideal distance is a balance between being far enough to be free of the traffic flow, the groundwater threat, the odors, and so on, and being close enough so that the hauling costs are not too great. We can imagine a citizen's indifference curves as a series of concentric circles, with the ideal circle perhaps 25 miles from his or her town.[3] As the distance to the site increases beyond this ideal, the site slowly becomes less attractive.

We took this collection of concentric indifference curves as a characterization of NIMBY preferences. In a simple version of the story, we imagined that each citizen has NIMBY preferences and each legislator reflects the preferences of his or her town. In the model for a vote between two candidate sites, legislators vote for the site located on the higher indifference circle. It is possible to consider all the candidate sites and all pairings of votes and thus to construct the "core," defined as the set of sites that are undefeated in vote pairings.

If the core is not empty, the likely prediction is that the selected site will be somewhere along the state boundary. If there is a great deal of cycling between possible solutions or if there is no core, an even more interesting prediction emerges, a prediction suggested by Ferejohn (1987) in another context. He suggested that when there is great instability in the voting, there will be an incentive for the legislators to restructure the institutional apparatus in order to avoid the indeterminacy.

There are several ways of breaking the symmetries causing the indeterminacy. The decision procedure can be delegated to legislative committees (in which case Ferejohn's voting model becomes a causal explanation of committee structures). In addition to this possibility, in our case, exclusionary criteria can be set up to narrow the list of candidate sites, thus narrowing the indeterminacy. An agency, such as the SWMC, can be given special powers to break the indeterminacy, and so on. In general, we can predict that asymmetrical institutional structures will be set up if there is too much indeterminacy in the legislature (and if the need to act is strong enough). We called this stylized story the "voting model." On a spectrum of unified action within the legislature, "tag the weakest player" is at one end and the "voting model" is at the other.

We identified a few more partial stories. One emphasized the growth of local veto power by the cities and towns and led to the prediction that there would be substantial growth in compensation ("local veto" story).

The important thing about these stories was that they were alternative causal explanations (they were not mutually exclusive, but they emphasized different things). They led to different predictions, and they suggested that different causal factors were pivotal. Thus the different stories, or causal models, would be likely to imply different policy actions to achieve the same policy goal. To undertake policy recommendations we needed some way to decide which story, or combination of stories, best described the "real game." We needed normative criteria to help identify the "best" explanatory (or positive) model.

In other words, the approach we were taking was in the spirit of Bayesian decision theory. In this approach it does not make sense to evaluate one possible explanation or model in isolation. Instead, the approach is to identify alternative candidates and then choose the best one, comparatively and under some explicitly defined criteria of "goodness." This contrasts, for example, with an approach often taken in classical statistical theory, where a null hypothesis is specified, but the alternative hypothesis is left almost entirely unspecified.

A Bayesian Approach to Uncertainty and Learning

We now had several candidate stories. I drew a picture of a box on the board, labeling it "the modeler's alternative worlds." As part of the box, I left some space labeled "terra incognita" for stories yet undiscovered. A little way off I put another less tidily enclosed figure, which I labeled "real world." Philosophers would no doubt recognize the picture as a correspondence view of the world.

Your problem, I told the class, is to decide which story (or model) best corresponds to the real world. More specifically, I want you to

1. draw your correspondence map with the possible explanatory models in the box;
2. express your assessment by assigning subjective probabilities to each of the possible models in the box;
3. think about what kinds of new information might lead you to revise your current probability assessments and by how much, and then seek the information and revise your current probability assessments;
4. think about how to discover new stories or models that would offer new candidate structures in terra incognita;
5. find new information that will lead us to reinterpret the existing stories ("tag the weakest player," and so on); and
6. do the above tasks interactively and simultaneously assess the models by how well they correspond, in their entirety, to the outside world.

In Bayesian terms tasks 1 and 2 define the state space and assign the priors. Task 3 is the standard one of assessing the conditional probabilities and Bayesian updating. Tasks 4 and 5 are invitations to find what I call "Bayesian surprises." (Perhaps 4 and 5 are a bit of a heresy to orthodox Bayesians. In these surprises, you do not update according to Bayes' theorem and pre-assigned conditional probabilities; instead you redefine the problem, restructuring the state space and existing "priors.") Task 6, along with tasks 4 and 5, means that the process is not a mechanical one of following Bayes' theorem but requires interpretive judgment all the way through.

Because I was emphasizing qualitative rather than quantitative analysis, this open and informal version of Bayesian inference suited my purposes. The approach is Bayesian in spirit in that there is no null hypothesis that is treated differently from the alternative hypotheses; instead the candidate hypotheses that describe the state space are treated on an equal footing.

Subjective probabilities are used throughout. There is at least an informal version of Bayesian updating and sequential learning.

As with means-ends rationality, this Bayesian approach to uncertainty and learning is broader than just the domain of policy analysis. It is similar to the experimental approach in science generally, where there is much opportunistic problem rethinking going on simultaneously with the experimental process itself.

Nonetheless, the approach sketched here differed from what the students were used to. Their pattern was to start with problem definition, then spend much time researching and describing the relevant legislation, regulations, history, and so on. In the process, policy recommendations would "intuitively" emerge (without explicit modeling or analysis). And then more time would be spent refining the recommendations and finding ways of implementing them. Thus the careful diagramming of the siting legislation was natural to them while the formalization and stylization of the competing possible causal stories was foreign.

The students were reluctant to make numerical probability assessments. They preferred to talk about contingencies or facts or more descriptive material. But they began to assign probabilities and think more specifically about what they needed to learn.

To develop alternative explanatory models and help resolve and assess the most important uncertainties, we made a list of people we wanted to interview. This exercise of forming a research agenda was more focused than the shopping list put together at the beginning of the course. In the next few weeks, the middle part of the course, the students interviewed about thirty people and reported what they learned.

We learned that the bill proposed by the Environmental Quality Commission was amended by the legislature in two important ways. The proposed bill had no Siting Board; the Siting Board, which must approve the SWMC's choices, was added by the legislature. Also the power of eminent domain was originally given to the operating agency (SWMC) in the original bill, but the legislature took this power from the SWMC and gave it to the governor. As one student put it, "We have been drastically underestimating the role of the legislature in our class discussions." On learning about the amendments, the students revised upward their probabilities of "hot potato" being the true story. "Tag the weakest player" seemed less plausible with these extra vetoes; moreover, the students thought that the governor's use of eminent domain would be more politically visible and less easily used in a crude way, compared with the SWMC's possible use of the tool. The students had trouble assessing the "voting model" and tended to leave it alone.

But if the siting decision was a hot potato, it seemed that the governor might be getting stuck with it by taking on eminent domain. We found out that for highway construction, where the power of eminent domain is routinely given, the power is usually lodged in the operating agency (which would correspond to the SWMC). Thus giving the power to the governor is unusual. A couple of the interviewees suggested that the governor was "crazy" to take on this power, which, they suggested, he would not even be able to use because it was so politically visible.

We began to wonder if, were the "hot potato" story true, it might be irrational for the governor to take on eminent domain. (We were told by several people that the governor had not only not opposed this change in the proposed bill but had "welcomed" it.) We began to think that if we could not find a plausible reason for the governor's wanting eminent domain, the "hot potato" story might not be so sensible after all. Recalling that Kiernan (and others by this time) had said that the governor had agreed to close the Johnston landfill when there were still thirty years of life left in it, we began to look for reasons why the governor might agree to closing Johnston in the first place.

Robert Benedict, director of the DEM, suggested three reasons for phasing out the Johnston site. First, it was not fair for there to be only one giant site and only one town bearing the burden. Second, the site would have to be closed someday, and it was prudent to start thinking about alternatives early (implicitly suggesting that the process would be very slow). And third, he said there was political pressure. Benedict said that he had not suggested to the governor, or others, that there was a potential groundwater contamination problem. (Even though the Johnston site is only six miles away from the Scituate Reservoir, from which half the state drinks, there appears to be good evidence that the two locations are on different groundwater systems.)

The explanation we most often heard for closing the site was the simplest: confrontational politics. "The governor was simply tired of dealing with aRusso." "aRusso put a lot of pressure on the haulers with the slowdowns and the haulers put pressure on the governor." In addition, we were told that it was environmentally and administratively risky to have just one big site. "Johnston is already a superfund site, and what would happen if it were closed down?"

If a gradual phasing out of the Johnston site was sensible on political grounds, and perhaps on environmental and equity grounds as well, eminent domain had to be part of the package. Otherwise, because of local vetoes through zoning, there would be no credible bill at all. When the legislature did not want to lodge eminent domain at the SWMC, the governor was willing to take it. We were also reminded that the siting decision was designed to come after the next election, and if eminent domain had to be used, it would come later too. The "hot potato" story seemed intact and, if anything, strengthened by what we had learned so far.

We talked to people in another of the possible veto agencies, the State Planning Council. We had thought of this agency as fulfilling a procedural role; it would ensure that the SWMC's choices and justifications conformed with the statewide plan. However, we were told that the council would have a much more activist role.[4] The council might add their own exclusionary criteria, evaluate the suitability criteria and possibly make substantive choices over the sites themselves. The people we interviewed pointed to some discretionary language in the Siting Act that suggested that the council could play a more activist role than we had envisioned.

This did not sound like the "hot potato" story, in which the legislature delegates to a lead agency and then puts procedural hoops in the hands of others. It sounded like a new story, one that might tell about how agencies like to exercise their powers and increase their size. Perhaps we had found a new island in terra incognita.

We began to wonder about who appointed people to the various councils and agencies. It turned out that the governor appointed many of them: he had appointed the director of the Department of Environmental Quality and many of the members of the State Planning Council and the Siting Board, as well as commissioners of the SWMC. Various members of the legislature appointed others. Could there be a rough analogy with the federal government? The initial mandate (in this case site selection) is given to the operating agency SWMC, which like EPA is partly an executive agency and partly autonomous and the home of many of the technical experts. And then, having only partial control of the technical operating agency, the governor, like the president, encourages the power of a praetoria guard, a small, staff-like agency that exercises not only veto power but substantive discretionary power through its threats of "remanding" decisions. In this analogy, the State Planning Council would be like the federal Office of Management and Budget.

If the analogy was valid, this was another Bayesian surprise, which suggested rethinking and restructuring the state space itself. The "praetorian guard" story was another version of the "hot potato" story, but instead of the legislature giving with one hand and taking with the other, it was the governor who did that. Of course, it is not a question of a single explanation—both delegation stories could be parts of the actual situation.

The idea of a unitary government "tagging the weakest player" had faded. But of course, even if a fragmented, decentralized, and cumbersome process finally made a site selection, it might pick a politically weak town. We learned that one town, Tiverton, feared that it might be so tagged. In preparation for this contingency, the town had already hired a prominent lawyer to fight the possible future decision.

In short, we found several plausible explanatory interpretations, but one thing rang like a constant through them all—the NIMBY preferences of individual citizens that were reflected in the conflicts among towns and between towns and the state government.

Normative Issues and Bayesian Decision Theory

In Bayesian decision theory, normative and positive ideas are distinct conceptually but are interwoven in application. The interaction is simple and direct. It arises through the idea of instrumental rationality. An individual's actions are good or bad (to the individual) depending on how well the consequences of the actions lead toward the individual's goals. The goals (incorporated in utility functions) lead to rankings of actions; both the goals and the inherited rankings are normative in character. But to rank actions one must assess the relationship between action and consequence (which requires understanding the institutional process and predicting others' actions) and this effort is causally explanatory (positive) in character.

I have just mentioned instrumental rationality at the level of individual choice. We can also think about instrumental rationality at the level of policy analysis. We can try to identify policy goals and then rank candidate recommendations on the basis of how far the consequences of the recommended actions go toward achieving the goals. At this more general level of analysis, we may ask how the whole system of incentives should be changed (how the game form should be redesigned).[5]

With this interweaving of normative and positive ideas in mind, let us go back to the beginning of the course. On that first day, I stated the "is" and "ought" (positive and normative) questions separately and put them on an equal footing. But the class soon spent most of its time on description, explanation, and prediction—positive questions. These questions seemed more practical and concrete. The normative questions were important but vaguely specified, even though we spent time on them and had the largest task group working on them. But now the course was more than half over, and we still had not done justice to the normative issues. I began to ask again for descriptions of the policy goals and the candidate list of possible recommendations that we should be discussing. The students who had had little previous exposure to policy analysis tended to think that even now identification and analysis of policy recommendations was premature. Recommendations should "emerge intuitively" from the research, and we should formalize them only at the very end. They should somehow fall out of the rest of the analysis. It was vaguely "cheating" and "prejudicial" to anticipate the recommendations before the end, just as it seemed (to the students) to be cheating to anticipate the result of a scientific experiment before observing its outcome.

All this, of course, is distinctly counter to the view of a Bayesian decision theorist. Such a person wants to know, as soon as possible, the possible alternative outcomes of an experiment and the importance of the possible mistakes in interpretating the outcome. Otherwise, how can she undertake "pre-posterior analysis" (analyze the value of information) and optimize the design of the experiment?

Similarly in a policy context, the decision theorist wants to know right away what things are up for decision (what is the candidate list of possible recommendations), how good or bad are the consequences that might result from the possible policy actions, and how costly might the various types of mistakes in inference and implementation be. Otherwise, how can the policy maker (aided by the decision theorist) improve the design of the existing institutions?

We discussed these questions with the help of a few readings, sympathetic and unsympathetic to decision theory. I suggested that at least part of the role of the scientists was to suggest which policy actions were feasible and worth worrying about (e.g. double liners for the landfills, choosing sites away from uncontaminated aquifers), and part of the role of the policy people was to identify the most important uncertainties to be resolved and help focus the research effort. While the scientists are primarily interested in truth and uncertainty and the policy people in value and cost, each helps define the scope of inquiry for the other, interactively and simultaneously.

My impression was that this Bayesian vision was foreign to the students. For example, they tended to think that recycling was a "good thing" to do as a moral duty, but they were unenthusiastic about justifying it on simple cost-saving grounds (although they sometimes did so, perhaps as a way of convincing others). In any case, they began to worry about how the roles of science and policy interacted. They began to think about how the "is" and "ought" questions might be separated conceptually but interact in an actual decision problem, like landfill siting. An example helped clarify the "policy analysis" view, even though it did nothing to resolve the tension between action as moral duty and action as instrumental means to a goal. The illustrative example began with the policy goal of making a car more fuel-efficient. Responsible governmental policy makers (Senate or EPA staff) go to the engineers and ask what needs to be learned to achieve the goal. But at the same time, engineers (in industry and universities), knowing how the motor works, may be able to identify policy options that policy makers would not think of by themselves but once identified would find important. In the Bayesian view, this policy process should be interactive. But the links between science and policy (uncertainty and valuation) are basically instrumental under the framework of value-of-information concepts.

Our reassessment of the compensation issue illustrates our thought about the interaction of the normative and positive dimensions. Of the three groups of technical people working on the siting problem and consulting for the SWMC, one, a group of University of Rhode Island economists, was assessing costs.[6] These economists were assessing environmental costs and surveying preferences (to help weigh "suitability" factors). Essentially they were describing NIMBY preferences and predicting the consequences of them. The economists were saying: "Given the amount of conflict built by NIMBY preferences into the institutional process the way it now stands, the machine is likely to seize up. If you, the policy makers, want to have a process that results in a selected site, you may want to take another look at compensation for the host community." To the economists, compensation is a "natural" way of turning a coercive, paralyzed situation into a voluntary agreement. Not only did the economists view the policy process instrumentally, they also viewed their role as one of indicating means to specified ends. The economists were like the engineers, telling the policy people how the motor works.

Suppose that "increased compensation" is put on the list of candidate policy recommendations. Is this change a good thing? Now we have a frankly normative question. How do we deal with it?

The students began by recalling why they had not taken the compensation issue seriously in the first place. First of all, they had viewed compensation as something like bribery. A system of compensation would encourage the weakest and poorest town to accept the site, because the town could be "bought off" at the lowest price. Compensation seemed unfair and normatively bad. Second, significantly increased compensation was not practical. The statute provided for a low level of compensation, at least at the beginning, and the people we interviewed were not talking about compensation as a serious problem. (Instead they talked about increased participation, hearings, arbitration, and over-riding the local veto with eminent domain.)

In contrast, some of the economists were saying that in the long run greatly increased compensation was "inevitable." Moreover, compensation should be made an integral part of the site selection process through some sort of negotiating process rather than tacked on at the end. We learned that amounts of compensation are increasing and compensation is becoming a more central issue. We discussed auction systems. An earlier version of the case study described by Rabe and Gillroy in this volume provided an example of a bargaining process. We found other cases where compensation was taking on an increasingly important role.

Of course, there is now compensation going to Johnston, some of it obtained long after the siting decision through confrontational bargaining. Compared with the present approach in Rhode Island, a move toward the bargaining process employed in the Alberta case (described by Rabe and Gillroy in this volume) would probably involve a higher cost of compensation. Presently the municipal tipping fee in Rhode Island is about $13/ton. Taking environmental costs and the costs of upgraded replacement facilities into account, this fee probably represents a large subsidy to the towns.[7] Rates in other areas are in the $50–$60 range (and higher for commercial waste). The present envisaged level of compensation is about $1/ton. As a rough guess, a move to a negotiating system might raise the compensation cost to $5/ton.[8] Thus the increased compensation package is relatively small compared with existing subsidies and costs.

As an empirical matter, it seems clear that compensation is becoming an increasingly important mechanism for "getting things done." But how do we decide when compensation is compensation and when it is bribery? We talked about the concept of informed consent. According to that concept, people agreeing to a compensation package must be informed about the risks and costs of the site. And to have consent, they must not be coerced. The students were concerned that in the siting of a landfill a form of economic coercion could arise if a town was very poor and had few options besides accepting the facility and the offered compensation.

We came to the view that compensation was not good or bad in itself but depended on the circumstances. Compensation was important not only empirically as a mechanism for "getting things done" but also for its normative-positive interaction. Several students stressed normative reasons why compensation packages should be bigger and a more integral part of the siting decision itself. They argued that if a town is to take a landfill site, it would be unfair not to compensate it up to the amount of the burden the landfill imposed. And as in the Alberta case, the process of bargaining over compensation early in the process might help to inform and empower the communities.

The Student Report

In putting together their report, the students settled on the following list of goals:

1. Managing the solid waste stream in an environmentally safe way
2. Managing it efficiently, in an economic sense
3. Designing a siting process that ends up in sites chosen
4. Developing a less adversarial process and one that is more efficient administratively
5. Providing equity to the host community by protecting it from environmental costs and risks and attempting to make it "whole"

Candidate policy actions included:

1. Reassigning the veto arrangements among the towns and between the towns and the state government
2. Changing the amount and timing of participation in the form of hearings and comment
3. Changing the opportunities for judicial review
4. Changing the technical characteristics of the facility
5. Altering the amount of compensation and its function in the site selection process
6. Altering the lexicographic process in its mixture of excludability and suitability criteria

Even stated in their rather vague, consensus form, these statements and the emphasis they are given in the report differ from what is found in the government reports we read. In the student report there was more explicit focus on the possible gridlock of the present process, more on compensation and "fair" bargaining, less on specific procedural issues (such as extending a hearing period from 40 days to 80 days) and more on the design of the institutional process taken as a whole.

Many of the students in this course participated in a summer workshop on the SWMC. As their project theme they chose user fees for solid waste pickup at the curbside, and one student who participated in both the course and the workshop chose to continue this work for his master's thesis.

Conclusion

The three core ideas of policy analysis—separation (and integration) of normative and positive, game theory, and a Bayesian approach to uncertainty and learning—provided a structure to the course. In doing so, they led to a research agenda and a framework of interpretation. This approach to policy analysis no doubt influenced our results—not just the framework of interpretation but the interpretation itself. I asked the students to think about, over the summer, how the approach influenced the substantive interpretation. Of course, the influence of the approach on the outcome is not just a phenomenon of the approach taken here but of any approach.

There were several differences between what we did and what the students expected at the beginning. In this class, students were encouraged to do more modeling, more stylizations of competing explanatory stories, to identify competing possible policy actions earlier, to make explicit their normative criteria earlier, and to have a larger section of analysis between the description of the problem and the policy recommendations.

If I had used a conventional cost-benefit analysis as the structure of the course, I probably would have seen more tension between the "policy analysis" approach and environmental values. (In our normative discussions students were suspicious of assumptions of commensurability and valuing "everything" under the "measuring rod of money.")

The three core ideas were easier to use (and accept?) on the more positive side of the analysis, where we were talking about explanation and prediction. The students' approach to the "is" question shifted ground during the course. Instead of focusing on the predicted result of a particular decision, they became more concerned with explaining the decision process as a whole. They wanted to know which causal story made more sense. Still better, they wanted to know how they could learn which causal story made more sense.

The three core ideas were harder to use on the more normative side, where we tried to specify goals and select policy recommendations. I take this to support Kelman's and Gillroy's views about conflicting values.

I think the students' approach to the "ought" question shifted ground as well. Instead of focusing on the goodness or badness of a particular decision, they came to be more concerned with the goodness or badness of the whole decision process. They wanted to know how to design an institutional process so that it would lead toward the broad policy goals more directly than would alternative designs.

The students shifted from an initial focus on particularistic solutions (e.g. extend the hearing period from 40 days to 80 days) toward more systemic solutions. For example, to the extent that NIMBY preferences generated conflict and the "hot potato" story made sense, there is a tendency to establish delegation processes with ambitious mandates, narrow procedures and high prospects of failure. Given this analysis, a systemic solution would be to diminish the level of conflict. One means of doing so is through compensation mechanisms.

During the course, there was a growing interest in "market-based" tools of implementation. Compensation is an example. Because the course was about site selection, we did not address directly the financing issues, but in the follow-on work there was a strong interest in user fees. My impression is that if a particular market-based tool of implementation looks like it could be effective, it may be explored and recommended. If so, this would represent a change from the last twenty years, during which time environmentally oriented students tended to dismiss market-based incentives.

Kelman suggests that people may hold values in both the domain of environmental values and the domain of economic values, that there is a tension between the two domains, and that people switch between the two in a sequential way. I observed a tension between the two domains, for example between action as a moral duty and as an instrumental means (also reflecting the divergence between Kantian and utilitarian orientations). Means-ends rationality was a central part of the approach to policy analysis, but I did not observe students switching in and out of a commitment to it in a sequential way. Instead, I saw them focusing it on the parts of the policy analysis where it was least threatening to environmental values and most practical. In the case of landfill siting, this focusing was toward the positive analysis and the implementation.

Notes

1. NIMBY is the acronym of "not in my back yard." NIMBY is formalized in what follows as a particular type of preference structure.

2. Discussion with Julie Roque, March 1990. See also Roque (1990).

3. Thus a special case of NIMBY preferences is that in which the ideal circle is infinitely far from the town and the utility function is cone-shaped. In this case, being removed from the environmental disamenities is always more important than the transportation costs.

4. Daniel Varian (secretary of the State Planning Council and director of the Department of Administration) and Sue Morrison (director of systems planning, Department of Administration) were interviewed.

5. Some game theorists accept the first form of instrumental rationality at the individual level but reject it at the second level of policy analysis and recommendation. Richard McKelvey, for example, argues that policy analysis and recommendation are just behaviors on the part of individual players to redesign the game form in ways favorable to them. If one accepts this view, one can think of policy analysis as a technician's job and ask the instrumentalist question: If there is a (partial) consensus as to a set of policy goals, what are the best policy actions to achieve them?

6. The two other groups: engineers were providing information about the technical design of the possible sites, and computer specialists were developing computer maps of locational features (wetlands, roads, and so on).

7. The low fee structure for the towns was negotiated in the Siting Act, largely by Ken Payne, to obtain the support of the towns for the act.

8. The actual cost of compensation in Johnston is probably considerably higher than $1/ton (taking into account new buy-outs and buffer zones). As another consideration, an auction process that led towns to compete with each other might lead to lower "equilibrium" compensation costs than would be suggested by the experience of Johnston bargaining individually, after the fact.

References

Albert, Jeff, et al. 1990. "Siting Landfills in Rhode Island: Past Present and Future." Center for Environmental Studies, Brown University.

Baxter, William. 1974. *People or Penguins: The Case for Optimal Pollution.* New York: Columbia University Press.

Ferejohn, John. 1987. "The Structure of Agency Decision Processes" in *Congress: Structure and Policy*. Cambridge: Cambridge University Press.

McCubbins, Mathew, and Talbot Page. 1987. "A Theory of Congressional Delegation" in *Congress: Structure and Policy*. Cambridge: Cambridge University Press.

Roque, Julie. 1990. "Scientific Methods and Policy Choices in Risk Analysis: The Regulation of Air Toxics in Rhode Island." Unpublished Ph.D. dissertation, Brown University.

9

Intrinsic Value and Public Policy Choice: The Alberta Case

Barry G. Rabe and John Martin Gillroy

The sight of angry and frustrated individuals at public meetings screaming at civic officials and demanding that a proposed waste dump, sludge site, incinerator or landfill be located somewhere—anywhere—other than their community is increasingly familiar to residents of Canada and the United States. The growing unwillingness of individuals to accept the economic and social costs expected to accompany such new facilities constitutes a growing problem. Solid waste, hazardous waste, and radioactive waste continue to pile up at alarming rates with ever-fewer safe storage or treatment facilities available, posing serious political, environmental, and public health problems for provinces, states, and urban areas.

This chapter contends that the *not in my back yard* (NIMBY) syndrome is not inevitable. NIMBY politics, in fact, can be better understood—and possibly transformed—through investigation of the theoretical assumptions of policy makers and the reality they are interpreting. In so doing, this chapter concentrates on the formal and substantive characterizations of the agents involved in NIMBY situations and the strategic interactions that repeatedly foil facility siting in North America. For the purposes of this chapter, it will be assumed that the facility to be sited can be considered a collective good, that the groups opposing one another can be considered rational, and that they are involved in strategic interaction in which the combination of their individual choices produces the resultant collective outcomes.

Specifically, we will examine the NIMBY problem as a strategic confrontation between the public as a group of citizens, in a specific political

Parts of this chapter are taken from Gillroy (1991).

jurisdiction, and the forces (government, industry, and so on) that are proposing the siting. Assuming that public policy is a product of the moral and strategic framework within which individual citizens exist and interact, the way in which a decision maker understands this interactive situation will determine his or her approach to the policy issue. Moreover, having the wrong theoretical (strategic and moral) representation of the empirical reality of agents, their preferences and goals will cause the policy process to be dysfunctional in solving the problem under consideration.

We will argue that although the NIMBY problem in most Canadian provinces and U.S. states appears to be, and is, handled by the decision maker as a classic confrontation of rational agents in a *collective goods/prisoner's dilemma* it is essentially a matter of moral agents confronting uncertainty within an *assurance game*, unable to protect the intrinsic value of their moral integrity and coordinate their behavior simultaneously. We will describe the NIMBY situation, as we contend it is commonly understood, and then suggest how the assurance problem is a better description of the NIMBY phenomenon.

This chapter will draw on comparative case study analyses of hazardous waste facility siting in Canada and the United States, which is dominated in both nations by subnational units of government. Despite the growing need for safe facilities for disposal and treatment, only one facility with the capacity for both organic and inorganic wastes and multiple methods of treatment and disposal has been opened in these nations in the last decade. The approach taken to site this facility, in Alberta, Canada, in the mid-1980s, differs markedly from the approaches of other states (including Florida, Massachusetts, Minnesota, Michigan, New Jersey, New York, and North Carolina) and other provinces (including British Columbia, Ontario, and Saskatchewan) examined in our study. In all of these cases, the NIMBY syndrome has thwarted all proposed facilities in recent years. The Alberta situation, therefore, affords a distinct empirical case for considering alternative theoretical assumptions about the NIMBY phenomenon.

Theoretical Representations and the Application of Game Theory

The representation of a policy problem can contain many components, but this chapter is concerned with two of the most basic: the strategic and moral reality of the human agent whose cooperation is the goal of the policy maker. These components are theoretical, pertaining to situations in general. They therefore must be used as a basis of analysis only when built upon by the "facts" of each particular case. The general theory, however, is meant to be applied as a foundation for all NIMBY situations regardless of what is being sited, who is involved, or where the siting is located.

Game theory can prove extremely useful in this process. It is a type of formal analysis that allows a theorist to examine the intricacies and results of the interaction of actors who are assumed to be capable of strategic rationality (Elster 1979, 16–18). It is an acknowledgement that individual choices are not made in isolation and that what results from one person's choice is dependent (to a greater or lesser extent) on the choices made simultaneously by others.

Game theory is used by some social scientists as a formal mathematical process that allows for the examination of how individual preferences are aggregated into collective outcomes (Axelrod 1984). There is, however, another role that can be played by game theory when it is used as "a descriptive framework for categorizing social interactions" (Hardin 1988, 32). It is in this latter guise that we will be using game theoretic analysis.

We will assume that the game under consideration includes three components essential to the policy maker's formal and substantive description of the situation. These three components are (1) the collective good being provided, (2) the assumed moral character and preferences of the contributing rational agents, and (3) the framework of strategic interaction that produces an outcome to the game.

Within the context of the NIMBY problem the collective good to be provided is a hazardous waste facility that must be sited in a Canadian province or an American state. Therefore in both game situations it will be assumed that the collective good is identical.[1] It is with the character of the agents and the framework of strategic interaction that the two-game situation and the resulting policy prescriptions differ.

NIMBY: The Prisoner's Dilemma

If there is anything that seems obvious about a NIMBY situation it is the virulent self-interest of the participants. Those who want to site the facility have problems that only the existence of the facility as a collective good will solve; those fighting the proposed site acknowledge in one breath that it is a necessary collective good and in the next that they do not want it near them (Gillroy and Shapiro 1986). From the standpoint of the actors in this confrontation, it would appear that their first preference is to obtain the advantage of the collective good without having to pay for it. Their second preference is for both actors to contribute, while the third preference is not to have the good at all. Lastly, each actor could pay for the good alone, while others free-ride on their cooperation.

The prisoner's dilemma is a strategic interaction between rational agents in which the symmetric payoff structure reveals a situation where the best option for the players to choose as individuals causes each to end up with what is irrational for them taken collectively. The prisoner's dilemma is an example of what Hardin calls "the back of the invisible hand"(1983, 6). Each person prosecuting his or her self-interest fails to provide for the collective interest.[2]

In the prisoner's dilemma as a collective goods game, the two actors (Row) and (Column) each have a binary choice to cooperate in the provision of the collective good or to defect from such cooperation. These two choices render four joint outcomes with individual payoffs that are represented in the four cells of the matrix (see Figure 9.1). Each agent is assumed to have the same ordinal preferences over these outcomes, which can be ordered as follows:

1. Best—*Unilateral defection*
2. Second—*Universal cooperation*
3. Third—*Universal defection*
4. Worst—*Unilateral cooperation*

From within this strategic context, as we can see looking at player one's payoffs in isolation, it is the *dominant strategy* for each player to defect from provision of the collective good, whatever choice the other makes (preference 1 is preferred to 2, while 3 is preferred to 4). With both agents playing their dominant strategy, the resulting payoffs will be (3, 3) with an equilibrium of universal defection. This equilibrium is neither pareto-optimal nor unstable; it poses a dilemma for the players as long as they independently prefer joint cooperation to joint defection but collectively accomplish the latter (Goodin 1976, 166-168).[3]

The NIMBY syndrome has increasingly come to be seen by policy analysts and policy makers as a form of prisoner's dilemma. States such as Massachusetts have viewed hazardous waste facility siting as a two-party conflict in which cooperation can emerge if negotiation is mandated between site proponent (usually a private corporation) and the local "host" community (Bacow and Milkey 1982; Holznagel 1986). Proponents of mandatory negotiation have expressed confidence that prolonged bargaining can lead to an iterative version of the prisoner's dilemma game, in which case an agreement becomes quite likely. In practice, this mandatory negotiation approach produces an interaction that more closely resembles a one-shot prisoner's dilemma game and ends in conflict (Rabe 1990). Widely heralded as a breakthrough when enacted in 1981, the Massachusetts approach has failed to attain a hazardous waste facility siting agreement in any of the six instances in which a developer has introduced a proposal. In an example of prisoner's dilemma logic, the potential host's preferred outcome is to thwart the

	C1	C2
R1	2 2	1 4
R2	4 1	3 3

2	4
1*	3*

Row Player's Payoffs in Isolation

*** Two Stars in Same Row = Dominant Strategy**

Figure 9.1 The Prisoner's Dilemma

proposed site; this may be realized if it resists ("defects") and the site developer withdraws ("cooperates"). Other options, in declining order of host preference, include compensated settlement between parties (universal cooperation), binding arbitration (universal defection), and, finally, developer success in imposing an agreement on the host (unilateral cooperation). In Massachusetts, the potential host repeatedly attains its preferred outcome, acting with such firm resolve that the developer withdraws its proposal. Five other states have modeled new hazardous waste facility siting on the Massachusetts approach, and none has attained a siting agreement to date.

Many policy analysts, approaching a siting policy through the framework of a prisoner's dilemma situation, have assumed that the prospective "hosts" of a facility desire to maximize their personal welfare within a preference schedule that would rather have use of the good without it being sited near them. The prospective "hosts" wish to "free-ride" on the cooperation of whatever other community, state, province, or nation gets stuck with the site. Under these circumstances, only two siting policies make sense: market-driven and regulatory-driven siting. The first policy is to leave the siting to the free market and allow rational agents to volunteer for a site if the project and its attached package of economic compensation is seen as ultimately beneficial to any community given the costs. Subnational governments play a passive role under this approach, whereas private corporations make crucial decisions on the types of facilities needed, where they should be located, and how agreements can be reached. There is no provision under this approach for governmental assistance in the bargaining process, as in the Massachusetts case. Such a market-driven approach reflects confidence that the two-party interaction will lead to a compensated settlement and universal cooperation. Numerous provinces and states—including British Columbia, Michigan, North Carolina, and Saskatchewan—have experimented with such a policy, none with siting success to date.

The second option is to coerce communities to accept sites through the regulatory tool of legal preemption. Anticipating local opposition to any proposed bargaining with a site proponent, states such as Florida, New

Jersey, and New York and provinces such as Ontario have attempted to impose sites on reluctant communities through preemption-type powers. However, local response has been so aggressive, in all such cases examined, that no new siting agreements, either in Canada or in the United States, have been reached under the regulatory-driven approach.

The prisoner's dilemma assumes an amoral, non-cooperative attitude on the part of the prospective "hosts." Therefore, the analyst could expect that the self-interested demeanor on the part of the rational agents involved will manifest itself in passionate efforts to see the site located somewhere else. A rational policy reaction to this would be to try to make as many decisions, short of the actual siting, in as much secrecy as possible. If it is assumed that the individuals involved will not contemplate cooperation in the provision of the good to be sited, then it is best for either the private developer or subnational agency to keep the plans away from a community until it has been decided that it is indeed the best "host" or is at least on a short list of acceptable sites. This understanding of siting politics is consistent with the guiding view behind the policy makers' assumptions in most provinces and states examined in this study and is evident under both market-driven and regulatory-driven strategies.

Overall, the prisoner's dilemma assumes a narrowly self-interested citizen who relies on cost-benefit calculation of his or her personal welfare as the basis of individual choice. If one assumes that the siting will always impose costs of one type or another that will not be overcome by welfare benefits, then cooperation in providing the site is very unlikely without coercion. Theoretically, of course, if the situation is a prisoner's dilemma with welfare efficiency as the major concern of the players, then compensation packages could be made large enough to gain prospective "hosts." Many subnational units, including British Columbia, Massachusetts, and North Carolina, have based their policies on this premise of potential compensation.

In this context, the main hope for cooperation is to iterate the game in order to facilitate bargaining over compensation and, more importantly, to evolve the norms of cooperation necessary to provide the collective good over a longer term without central coercion (Hardin 1983). In this way a normative character can be brought into the game from outside and a "moral" connotation given to the cooperation of the players so that siting decisions may be made more acceptable to all involved. However, the dominant strategy of the player remains to defect from provision of the good. Even in iteration with the evolution of cooperative or "moral" norms, one thing remains true: if the cooperative equilibrium is established, no one agent will make the first move toward it, while it is in every agent's interest to take the first step away, cooperative norms or not.

Under these circumstances, the seeming risk averseness of the agents can be chalked up as the manifestations of narrow self-interest in the face of a site that will have more welfare costs than benefits to the surrounding inhabitants. If this is the only way to characterize the hazardous waste facility siting process, the NIMBY syndrome may only grow stronger in the years ahead, leaving waste exports and illegal dumping as increasingly common policy options because old facilities will continue to close and new ones cannot be opened or even sited as replacements.

NIMBY: The Assurance Game

There may be, however, at least one alternative explanation for the behavior of agents in a NIMBY situation. If applied to hazardous waste policy, it could foster new—and more promising—approaches to facility siting. This alternative is the strategic framework of an assurance game (Sen 1967; Elster 1979, 20–21). The first major distinction of an assurance game is that the preferences of the agents define the best outcome in terms of cooperation in provision of the collective good. (See Figure 9.2.) The ordinal preferences of each player are:

1. Best—*Universal cooperation*
2. Second—*Unilateral defection*
3. Third—*Universal defection*
4. Worst—*Unilateral cooperation*

Within the assurance game, it is assumed that the ordinal preferences of the players are arranged so that the best and second-best preferences of the agent in the prisoner's dilemma are reversed. Thus the best preference is universal cooperation and unilateral defection is only second best. Unlike the prisoner's dilemma, where one will cooperate only if he or she cannot free-ride, a player in an assurance game will only free-ride if he or she cannot cooperate.

The logic here describes a rationality of intention to cooperate with an underlying capacity, and ability, to exploit. Unlike the prisoner's dilemma where exploitation of others is the best alternative for the player, in the assurance game the cooperative solution is preferred but the ability to exploit looms as a pertinent consideration of each player's strategic rationality. The worst result for each player is still exploitation in unilateral cooperation, and it is still the case that even a failure to coordinate at all is a better option than being exploited in this way.

However, a significant difference between these games is that whereas information and central assurance that all will cooperate are neither necessary nor sufficient to establish a stable cooperative outcome in the prisoner's dilemma (for each player can be expected to free-ride whenever possible), they are both necessary and sufficient for the same ends within the assurance game. In fact, within the assurance game, without a commitment to broad dissemination of accurate information, open communication, and central regulatory involvement that credibly assures all players that they will not be exploited in the decision to cooperate and act as moral persons, the game can degenerate into a fight for survival and a ruination of collective action. In such cases, each player, fearing that he or she will be exploited, moves to exploit the other in what Immanuel Kant called the "mania for domination" (Kant 1800, 273). Kant described the moral psychology of the assurance game:

> The citizen's inclination to do violence to one another is counteracted by a more powerful force—that of government. This not only gives the whole a veneer of morality . . . but by putting an end to outbreaks of lawless proclivities, it genuinely makes it much easier for the moral capacities of persons to develop into an immediate respect for right. For each person believes of himself that he would truly venerate and abide by the concept of right, if only he could expect the same from everyone else, which it is government's part to insure. . . . But since all persons have a good opinion of themselves but presuppose evil intentions in everyone else, they mutually have this opinion of one another, that they are all, in point of fact, of little worth. (Kant 1795, 375 fn.)

The precarious nature of the intention to cooperate and the critical nature of outside signals to the individuals playing the game can be seen in the fact that the assurance game has no dominant strategy (see player one's isolated payoffs—Figure 9.2). If one is assured that the other player will cooperate, then one prefers to cooperate. However, if no such assurance is forthcoming, then one prefers to defect and protect oneself. Unlike a prisoner's dilemma where each player prefers to defect whatever the other players do, the lack of a dominant strategy in an assurance game makes one player's payoffs contingent on the decisions of others or the assurances provided by third parties. The uncertainty of this game, lacking a dominant strategy, simultaneously makes the role of central coordination (public policy) very important and the specter of the "mania for domination" simultaneously more real and less inevitable than in the prisoner's dilemma.

Within the confines of NIMBY-related policy, the analyst who sees the situation as an assurance game has a completely different set of theoretical assumptions upon which to devise policy. The character of the agents is richer in that a distinct sense of the collective interest becomes the first

	C1	C2
R1	1, 1	4, 2
R2	2, 4	3, 3

	C1	C2
R1	1*	4
R2	2	3*

Row Player's Payoffs in Isolation

*** Stars in Different Rows = No Dominant Strategy**

Figure 9.2 The Assurance Game

priority for each person. There now exists an internal "moral" recognition, in the rationality of the player, that one's self-interest involves the provision of collective goods and cooperation to solve collective action problems. This elevates the strategic situation from one of narrow calculation of costs and benefits to wider consideration of the necessity of a "collective good" and its relation to the essential nature of the individual as a cooperative moral agent. For the first time, the citizen or policy maker might recognize that the ethical consideration informing his or her deliberations is not the instrumental value of efficiency but the intrinsic value of individual moral integrity.

> Public policy, then, may not represent an attempt to increase welfare, utility or "satisfaction." It may represent our attempt, rather, to control the conditions under which we pursue happiness—the conditions under which we lead our lives. (Sagoff 1982, 761)

If this is the case, then the passionate self-interest that is manifest in the NIMBY syndrome may be an artifact of fear that one may lose moral integrity by unilaterally cooperating without assurance that one will not be exploited. That is, one may wish to cooperate, to contribute to the provision of the collective good. But without assurance that one's participation will not be exploited by other agents, one will retreat into more narrowly self-interested behavior in self-defense, triggering Kant's "mania for domination." It is no longer a case of accepting a bad payoff that does not pass a cost-benefit test, but the possibility of having one's moral integrity, or intrinsic values as a moral agent, exploited by others.[4]

Within the assurance game, norms of cooperative behavior are not evolved externally, as in the iterated prisoner's dilemma, but instead pre-exist, internal to the game, as the moral predisposition of each player to express his or her rational agency in choosing to cooperate. The task for the coordinator is not to set up a system of norms but instead to recognize the norms inherent in protecting and empowering the humanity of each person and their integrity as a moral agent.[5] Therefore, within the assurance framework the policy maker finds individuals who are predisposed to be moral, with the assurance from government that their autonomous actions will not be exploited. This establishes an active role for the policy maker as representative of the citizens involved in any particular siting and suggests certain political actions as necessary to preventing the NIMBY syndrome.

Toward the Assurance Game: The Case of Alberta

The case of hazardous waste facility siting in the Canadian province of Alberta indicates that it may be possible to build siting policy on the theoretical foundation of an assurance game setting. The new facility that opened in Alberta in the late 1980s constitutes the first step toward a comprehensive Canadian system of hazardous waste management. It is intended to employ more sophisticated technologies for waste treatment and disposal than the landfills, incinerators and open dumps that were scattered across Canada (many of which continue to operate) and managed in slipshod fashion in prior decades. The siting agreement emerged from a provincial siting process that differs markedly from those used in other provinces and states.

The province of Alberta opened the Swan Hills Special Waste Treatment Center in September 1987, following a radical transformation of the provincial siting process earlier in the decade. Site proposals advanced by private corporations in the late 1970s met with outraged local resistance in the communities of Fort Saskatchewan and Two Hills, and the prospects for any future agreement seemed limited at best. In response, the Alberta environment ministry declared a moratorium on future siting and established a committee to review the provincial siting process and propose reforms.

This resulted in a new process that involved the province as the central coordinator in all siting situations, contrary to earlier policy whereby private corporations dominated the process. Alberta established a Crown corporation to manage any major waste facility jointly with a private corporation that would be selected by the province. It also developed a comprehensive program for educating the public and considered siting only in communities that met provincial environmental standards and volunteered as site candidates. After a multi-layered process of consultation with varying local government jurisdictions, five communities came forward as potential site hosts. Swan Hills, a community of 2,000 residents located 209 kilometers northwest of Edmonton, was selected in 1984 after a plebiscite in which 69 percent of its voters supported the facility proposal.

The Waste Treatment Center is the most comprehensive waste disposal facility in either Canada or the United States, with the capacity to incinerate organic liquids and solids, treat inorganic liquids and solids, and landfill contaminated bulk solids. It has a potential capacity of more than 100,000 metric tons of hazardous waste annually, adequate to cover the vast majority of hazardous wastes generated in Alberta each year. It is designed as part of a comprehensive waste management program, which includes regional transfer stations and an extensive program that promotes waste recycling and on-site waste reduction.

The following case analysis suggests that a number of components of the Alberta approach to siting were distinctive from those in other provinces and states and that these served to transform the siting process from the common prisoner's dilemma format to the assurance game framework. Economic compensation, it will be argued, was not inconsequential to the agreement but was influential only after being placed in a distinctive context that created an assurance game setting and guaranteed ex ante protection of the moral integrity of the agents involved as a precondition of their cooperation in the siting. The following factors proved central to this transformation.

Tripartite Management and Governance

New, intermediary institutions have often served to transform highly conflictual situations into more cooperative ones by providing safeguards against exploitation. For example, Robert Keohane argues that new multinational institutions have played a pivotal role in fostering cooperation in an era in which no single nation or institution is likely to enforce agreement through hegemonic power (Keohane 1984). New institutions may be needed to transform the siting process. These institutions will be distinct from traditional agencies within individual states and provinces that lack public credibility, meet stiff resistance and repeatedly fail to attain agreement. The introduction of a Crown corporation into provincial hazardous waste management appears to have contributed to the cooperative outcome in Alberta. This corporation assumes a number of the important responsibilities delegated to either private developers or regulatory agencies in most states and provinces (Laux and Molot 1988). It provides for direct governmental oversight of facility operation and also affords uniquely direct public financial and technical assistance to private corporations responsible for site development and management. As a 1981 government report endorsing the Crown corporation concept noted, the corporation

> would provide effective evidence of an arm's length position relative to government and industry . . . while allowing various government departments to continue their particular regulating, inspecting and monitoring functions. . . . The public would be more likely to trust the administration of

> a crown body. Industry has indicated that if allowed to operate facilities in a free market environment, they too could function efficiently under such administration. Therefore, both concerns are met. (Alberta Hazardous Waste Team 1981)

Under this tripartite system, the Alberta Special Waste Management Corporation is responsible for overseeing numerous aspects of the provincial waste management system, including plant design and construction, provision of 40 percent of construction and operating costs plus operating loss subsidies to the private corporation, control of all provincial transfer and collection points, collection of 40 percent of revenues generated by the facility, provision of utilities and highways for the facility, and ongoing research, monitoring, and technological appraisal. It also owns the site, which it leases to the private firm for a minimal fee. In turn, Bow Valley Resource Services provides 60 percent of the construction funds and operating costs and handles day-to-day operation of the facility.

The Crown corporation is distinct from Alberta Environment and related provincial agencies, which set regulatory standards that specify the ways in which respective wastes are to be treated. Alberta Environment also provides a system for registering these wastes and punishing regulatory noncompliance by either the Crown or private corporations. At the same time that the Swan Hills siting decision and Bow Valley Resource Services selection were made, Alberta was devising one of the more comprehensive hazardous waste regulatory systems of all the Canadian provinces, which was to be implemented by these agencies.

Siting as Part of a Comprehensive Hazardous Waste Strategy

Hazardous waste facility siting is frequently presented as a one-shot, take-it-or-leave-it proposition from a private firm or subnational government to an individual community. In most provinces and states, it is not treated as part of a broader strategy to protect the environment and safeguard public health that might involve multiple communities simultaneously. The prospects for cooperation in Alberta may have been enhanced by taking such a larger perspective. In particular, Alberta officials argued convincingly that prior hazardous waste disposal was so widely and haphazardly done that a shift toward more sophisticated disposal and treatment was urgently needed. They also contended that acceptance of a major waste facility would be only part of a comprehensive provincial strategy; other communities would be expected to play an active role in the waste management process through acceptance of transfer and collection stations as well as involvement in extensive waste reduction and recycling programs. Finally, it was strongly implied in

Alberta that acceptance of a major facility would ultimately enable an individual province to take control of its own hazardous waste problem, excluding wastes generated in other provinces and ending the need to export wastes elsewhere.

Evidence was increasingly abundant by the early 1980s that unsafe waste dumping was prevalent and that no community could completely hide from potential toxic exposure. The Environment Council of Alberta found that "disposal of solid and liquid wastes on land is largely uncontrolled by legislation. Large volumes of the solids, sludges and liquids are put directly into landfills which are not suitable for them" (Environment Council of Alberta 1980). Such revelations hit close to home in Swan Hills, where slipshod disposal practices were commonplace, particularly for wastes from the oil and gas extraction industries (Bohn 1986). A government report revealed that hazardous waste disposal had been "left to the generators of the waste. It is routinely concealed as normal waste and disposed of in unsuitable areas" (Alberta Hazardous Waste Management Committee 1979).

Making the case that there was a broad environmental and public health threat, was easier, however, than devising a comprehensive approach to waste disposal that involved multiple communities and industries, and, in effect, the entire citizenry. Nonetheless, Alberta officials have been successful in addressing provincial hazardous waste policy as more than the mere construction of a new disposal facility. This has helped create a collective sense of responsibility for waste management rather than deposit all the costs of disposal on a single community. Alberta proposed a new program for registering wastes and tracking their transfer and called for a system of regional facilities to accept hazardous wastes from surrounding areas and prepare them for transfer to the central provincial facility. The province also devised ambitious programs to promote waste reduction and recycling by providing incentives to reduce the volumes of wastes that would be sent to Swan Hills and creating a waste exchange system whereby one firm's wastes might be reused in the production process of another.

Provincial officials also attempted to build support for new facilities by portraying them as a linchpin in provincial plans to take full responsibility for only those wastes generated within provincial boundaries. On the one hand, officials lamented the growing reliance on exporting hazardous waste to other provinces and states as both morally offensive and economically risky given the uncertain future availability of disposal facilities elsewhere. Consistent with the statements of others, a senior provincial environmental official expressed disapproval of the Albertan reliance on Oregon facilities to dispose of about 1,000 cubic feet of PCBs per month. He noted: "I feel it's irresponsible for us to send our wastes across the border. . . . And I would be pleased to see the [U.S. Environmental Protection Agency] stop the practice" (Morningstar 1979). His suggested alternative was to proceed with the business of developing a major disposal facility and comprehensive hazardous waste policy in the province, thereby precluding the need for additional exports.

On the other hand, officials vowed to attempt to confine use of new waste disposal facilities, if accepted, to those wastes generated in the province. Alberta officials vigorously resisted efforts in 1981 to explore regional disposal plans that called for housing the main multi-provincial facility in their province. They frequently stated their intent to minimize waste trading with other states and provinces upon opening a comprehensive facility in Alberta.

Recognizing Governing Norms

Norms that guide behavior and lead to a collective willingness to address the problem of hazardous waste management have been notably lacking in most state and provincial siting efforts. A norm functions in a particular social setting insofar as individuals agree to act in certain ways and be punished when they fail to act in these ways. Norms may be buttressed by laws but take on a self-policing characteristic that is often fundamental to cooperative interaction and implementable policy (Axelrod 1986). With regard to hazardous waste, neither Canada nor the United States has devised generally acceptable understandings of what constitutes appropriate and inappropriate conduct either individually or by private or governmental organizations. This has led to extensive illegal dumping in both nations and explains the proliferation of abandoned sites that has necessitated, in the United States, the creation of a multi-billion dollar Super Fund to facilitate highly expensive site cleanup. It has also encouraged exportation of wastes to developing nations, which has mired both countries in embarrassing foreign policy conflicts upon revelation of haphazard dumping in heavily-populated areas abroad.

Alberta has not resolved this issue but has taken unusual steps to begin to develop governing norms by recognition that external "policy norms" must be based on existing moral priorities and values that the population takes into any collective action situation. Alberta, through massive information and educational efforts, has sought to recognize the shared values underlying the predisposition to cooperate and has respected the fundamental capacities of its citizens to exhibit their integrity as moral and political agents by seriously including their input at all points in the process. "The public needed to be able to identify with the problem before it even considered taking any responsibility in developing a solution," noted one analyst (Morningstar 1979). These efforts are intended eventually to provide an over-arching framework in which all Alberta citizens can begin to understand their personal contributions to the problem and their potential role in its resolution. These efforts have only begun to lead to norm development that might facilitate a collective sense of responsibility for the hazardous waste problem. But they appear to surpass those that have been attempted in the other provinces and states that were examined, increasing the likelihood that siting can proceed in the context of an assurance game.

Early and Extensive Public Involvement

Provinces and states have consistently approached the question of public participation in various stages of the siting process with pause. Many subnational units have delayed public involvement until the later stages of the process, after a specific site has been announced by either a governmental agency or a private corporation. This may well be too late to attain meaningful public input or to build trust between the site proponent and the proposed host. As Bruce Williams and Albert Matheny have noted in their examination of public participation in siting among the American states, "the systematic and sustained effort necessary to create true democratic dialogue among citizens" has been sorely lacking (Williams and Matheny forthcoming).

Such effort, in contrast, has been unusually extensive and successful in Alberta. The committee assigned to devise a new provincial siting process was emphatic about the need for extensive public participation measures that would educate the public on the nature of the hazardous waste problem and various disposal options, fully involve communities that were potential site candidates, and assure all provincial residents that any final decision on a site would require formal indication of local citizen support. According to one Edmonton newspaper columnist, such provisions for public input into provincial decision making were "almost revolutionary in Alberta" (Livingstone 1981, A6).

These participatory efforts were also time-consuming and involved extensive provincial commitment of staff to travel and visit various community groups. Alberta Environment officials hosted more than 120 meetings in every county, municipal district, and special area in the province. They responded to citizen inquiries, provided briefings on the hazardous waste situation in the province, and offered information on the procedures to be followed in the siting process (McQuaid-Cook and Simpson 1986, 1031–1036). Fifty-two of a possible seventy communities expressed interest in potential participation and continued to have far-reaching access to provincial officials and hazardous waste data. They also received a detailed provincial analysis of their area, which would prove useful to them in considering the viability of a hazardous waste site as well as landfill sites or other land uses.

Fourteen of these communities sought further consideration by the province, although nine were subsequently eliminated either on environmental grounds or in response to vocal public opposition. The five remaining communities held plebiscites in 1982 that drew heavy voter turnout and support for the idea of hosting a hazardous waste facility. This gave provincial officials a choice among possible sites, leading to the selection of Swan Hills in March 1984. Substantial public participation has continued since the opening of the facility through a number of formal and informal mechanisms that maintain communication between Swan Hills' residents, provincial authorities, and representatives of the Crown and private corporations that manage the facility.

Compensation

The practice of placing primary reliance on compensation to resolve siting conflicts has failed dismally both in Canada and in the United States. The idea of offering compensation has failed to meet expectations that it could foster universal cooperation in a prisoner's dilemma game. However, compensation can play an important role in promoting siting agreements when packaged as part of a larger, more structured process, after the moral integrity of the agents has been assured. In such a context, the anticipated economic—and social—compensatory benefits were significant contributors to the siting agreement reached in Alberta.

The extensive participatory processes in Alberta made possible prolonged bargaining between province, private corporation, and local officials concerning the compensation benefits that would be accorded Swan Hills in the event that the community accepted a comprehensive hazardous waste facility. Both the Alberta Crown corporation and privately held Bow Valley Resource Services provided numerous incentives to cooperate. These included grants for a golf course and other recreational development, funding to hire consultants to scrutinize the proposal and eventually monitor the facility, construction of subsidized housing of approximately thirty-five family units, purchase of a van to transport local residents to the site, and making headquarter offices available for public meetings, among many others. The facility has led to the creation of eighty-six new jobs, the vast majority of which are held by Swan Hills residents. Furthermore, local officials remain convinced that the construction and opening of the facility provided a stimulus for considerable new development, including the opening of a new hospital and new office complex and planning for a major industrial park. All of these were attractive to Swan Hills residents, who suffered in the late 1970s and early

1980s after the collapse of the oil and gas extraction industries, as did citizens of many other western Canadian towns of the period (Williams 1978; McParland 1981).

NIMBY: The Policy Imperatives

The Alberta case suggests that subnational governments must not only protect the autonomous choices of their citizens but must also encourage and facilitate their participation in every aspect of the siting process. This requires that the government act as full partner in every aspect of the siting process and ultimate facility operation, providing constant assurance against the threat of exploitation. Individuals must be able to trust the government as a representative of the collective interest in the provision of collective or public good if siting is to occur in an assurance game context.

An important difference between the prisoner's dilemma and the assurance game, and therefore an important foundation in policy formulation, is the role of full information in the assurance game.

> In the prisoner's dilemma the requirement of full information is not necessary for the solution to emerge, but in the assurance game the slightest uncertainty or suspicion will make an actor choose [to defect] rather than [cooperate]. Only in a fully transparent situation will the actors converge upon the collectively rational behavior. Given perfect [trusted] information, the difference between the two games may be expressed by saying that in the prisoner's dilemma the [collective] optimum is both individually inaccessible (no one will take the first step toward it) and individually unstable (everyone will take the first step away from it), whereas in the assurance game while it is still individually inaccessible the solution, once reached, is individually stable. (Elster 1979, 21)

Before the actual siting takes place it is incumbent on the government to begin a dialogue with its citizens about the general parameters of the problem that makes the siting necessary, the dynamics of the problem, and the range of possible solutions (Williams and Matheny forthcoming). Only after full participation and explanation of the necessity and risks of the siting is it appropriate to consider potential hosts. The assumption within the assurance problem is that individuals are not necessarily risk averse but want to be respected in their choice to accept a risk to themselves and their community. This respect for their moral integrity requires full disclosure. Such disclosure—and systematic roles for public input—was crucial in the Alberta case. Disclosure has been largely absent in virtually all of the other provinces and states examined in this study.

Each community must believe that it is requesting the site as its contribution to the collective good with the full assurance that it will be protected by its government regulators from exploitation by the site development or the externalities of the resulting facility. Throughout the process it must be made evident that the site is a collective good necessary to the citizenry and this must be argued to the satisfaction of those involved before any sites are chosen, even on a preliminary basis. It must also be clear that the construction and operation of the facility is only part of a comprehensive provincial, state, or regional strategy and that the host community is not alone in working toward hazardous waste policy that will protect the environment and safeguard public health. Any new facility must be linked to far-reaching waste recycling and waste reduction initiatives as well as regional facilities for sorting wastes and preparing them for transfer to the disposal facility. This, too, was a crucial component in the siting policies that led to an agreement in Alberta.

The collective good must be seen as a public trust protected by government on behalf of the citizens, who have the decisive power to allow siting only when done in the public interest. Here the benefits of a site are not solely economic and social compensation but also acknowledgement of the "public spirit" of a community. Each site is agreed to with full knowledge of its importance and its risks but with the assurance that "free-riders" will be controlled. This requires that the government representatives have pre-existing regulatory powers to monitor the site on a continuing basis and to control waste exports to, and imports from, other subnational units of government. This involvement need not take the formal shape of a Crown corporation, although some institutions are necessary to assure a substantial—and enduring—governmental role.

If citizens in an assurance game are confident that their subnational government represents the public interest in providing necessary collective goods and they are allowed full participation in the myriad decisions that must be taken before a facility type or site is chosen, then their intrinsic value as moral agents will not be jeopardized. Moreover, their rationality will be free to function without the pressure of fear and uncertainty. They will be able to approach the problem as a collective one in which they are intimately involved in the decisions and privy to all information necessary to make them. Recent experience in hazardous waste facility siting in Canada and the United States suggests that this may be the only way to begin to move beyond siting gridlock and toward a more mature system of waste management.

Notes

1. This is not to say that the degree of tension will not vary with the specific facility being sited and its level of risk. For the purposes of this chapter, however,

we will assume that the siting problem is generic and identical strategically for each site within each game.

2. In the original context of the prisoner's dilemma, two criminals have been arrested. The prosecutor has sufficient evidence to convict both of them on a minor charge; he also suspects that they have committed a serious offense, but he does not have enough evidence to convict them of it unless one confesses. The prosecutor approaches each prisoner with a deal and he tells each that he is offering the same deal to the other. The deal is this: (1) if neither prisoner talks, conviction on the minor charge is certain; (2) if one prisoner turns state's evidence unilaterally, implicating the other in the serious crime, he will go free; (3) if the other prisoner implicates him unilaterally, the prisoner will receive harsh punishment on both charges; (4) if both confess, both will receive more punishment than they would for the minor charge but not as much as they would if they were punished on both charges.

3. Cooperation can be established from within a prisoner's dilemma by iterating the game. However, this cooperation is ineffective without some kind of external constraints (norms, conventions, law), while the primary preference of each player is to take advantage of the other (Hardin 1983, Chapters 10, 11, 12; Axelrod 1984).

4. Thomas Hobbes understood the contention that to hold a moral predisposition is different from being able to act upon it. "The Lawes of Nature oblige in foro interno: that is to say they bind to a desire they should take place: but in foro externo; that is, to the putting them in act, not always. For he that should be modest, and tractable, and performe all he promises, in such time, and place, where no man els should do so, should but make himselfe a prey to others, and produce his own certain ruine, contrary to the ground of all Lawes of Nature, which tend to Natures preservation" (Hobbes 1651, 215).

5. This does not necessitate replacing a concept of self-interest with one of altruism. Kant understood the evolution of one's self-interest to begin with one's animal nature and end with "rational self-love," which can be defined as the capacity to see one's own real interests in protecting and empowering one's capacity as a human agent. This entails a further "moral" recognition that cooperation with others as ends-in-themselves is necessary to one's self-interest (Kant 1788, 73).

References

Alberta Hazardous Waste Management Committee. 1979. *Reid Crowther and Partners Report: A Summary of Inventory and Present Disposal Practices.* Edmonton: Alberta Hazardous Waste Management Committee.

Alberta Hazardous Waste Team. 1981. *Hazardous Wastes in Alberta.* Edmonton: Alberta Environment.

Axelrod, Robert. 1984. *The Evolution of Cooperation.* New York: Basic Books.

————. 1986. "An Evolutionary Approach to Norms." *American Political Science Review* 80:1095–1111.

Bacow, Lawrence D., and James R. Milkey. 1982. "Overcoming Opposition to

Hazardous Waste Facilities: The Massachusetts Approach." *Harvard Environmental Law Review* 6:265–305.

Bohn, Glenn. 1986. "Where Waste Finds a Home. . . ." *Vancouver Sun*, 22 November, p. B1.

Elster, Jon. 1979. *Ulysses and the Sirens: Studies in Rationality and Irrationality.* Cambridge: Cambridge University Press.

Environment Council of Alberta. 1980. *Hazardous Waste Management in Alberta.* Edmonton: Environment Council of Alberta.

Gillroy, John Martin. 1991. "Individual Autonomy and the NIMBY Problem" *Public Affairs Quarterly* 5:319–332.

———, and Robert Y. Shapiro. 1986. "The Polls: Environmental Policy." *Public Opinion Quarterly* 50:270–279.

Glenn, William M., Deborah Orchard, and Thia M. Sterling. 1988. *Hazardous Waste Management Handbook.* 5th Edition. Don Mills, Ontario: Southham.

Goodin, Robert. 1976. *The Politics of Rational Man.* London: John Wiley.

Hardin, Russell. 1983. *Collective Action.* Baltimore: Johns Hopkins University Press.

———. 1988. *Morality Within the Limits of Reason.* Chicago: University of Chicago Press.

Hobbes, Thomas. 1651. *Leviathan.* C. B. Macpherson (ed.). Pelican Classics Edition. Harmondsworth, Middlesex: Penguin Books.

Holznagel, B. 1986. "Negotiation and Mediation: The Newest Approach to Hazardous Waste Facility Siting." *Boston College Environmental Affairs Law Review* 13:327–378.

Kant, Immanuel. *Gesammelte Schriften.* Berlin: Prussian Academy. 1788. *Critique of Practical Reason.* 1795. *Perpetual Peace.* 1800. *The Anthropology.*

Keohane, Robert O. 1984. *After Hegemony.* Princeton: Princeton University Press.

Kuflik, Arthur. 1984. "The Inalienability of Autonomy." *Philosophy and Public Affairs* 13:271–298.

Laux, Jeanne Kirk, and Maureen Apell Molot. 1988. *State Capitalism: Public Enterprise in Canada.* Ithaca, N.Y.: Cornell University Press.

Livingstone, B. 1981. "Hazardous-Waste Plants Needed." *Edmonton Journal*, 28 September, p. A6.

McParland, Kelly. 1981. "Oil Cuts Drain Town's Lifeblood." *Edmonton Journal*, 10 July, p. B1.

McQuaid-Cook, Jennifer, and Kenneth J. Simpson. 1986. "Siting a Fully Integrated Waste Management Facility." *Journal of the Air Pollution Control Association* 34 (September):1031–1036.

Majone, Giandomenico. 1989. *Evidence, Argument, & Persuasion in the Policy Process.* New Haven, Conn.: Yale University Press.

Morningstar, Lasha. 1979. "Dumping Our Hazardous Waste Outside Alberta Irresponsible." *Edmonton Journal*, 22 September, p. A7.

Rabe, Barry G. 1990. "The Hazards of Institutionalizing Mediation in the Siting of Hazardous Waste Facilities." In *Conflict Resolution and Public Policy*, Miriam K. Mills, ed. New York: Greenwood. pp. 3–20.

Sagoff, Mark. 1982. "On Markets for Risk." *The Maryland Law Review* 41:755–773.

Sen, Amartya. 1967. "Isolation, Assurance and the Social Rate of Discount." *Quarterly Journal of Economics* 81:112–124.

Williams, Bruce, and Albert Matheny. Forthcoming. *Democracy, Dialogue, and Social Regulation: Being Fair Versus Being Right.*

Williams, Ian. 1978. "Swan Hills Still Carries Transient Tag." *Edmonton Journal*, 4 January, p. B1.

Epilogue: Environmental Values and Economic Trade-offs—Conflict and Compromise

Douglas MacLean

In his campaign speeches, George Bush, an oil state Republican, claims to be the champion of clean air and water, and he appoints a well-known leader of the environmental movement to be the administrator of the Environmental Protection Agency. Editorials calling for tougher fuel standards to fight pollution appear in *Road & Track* magazine. Industry giants like McDonald's join hands with the Sierra Club to develop environmentally benign packaging and to reap the benefits of appearing to be concerned corporate citizens. In households across America, environmentally conscientious citizens diligently sort and recycle their trash. There is a political and moral consensus today that we need to protect the planet. Everyone is an environmentalist. We can see it in opinion polls and in college course catalogs. Corporate leaders, the media, politicians, and their "handlers" are all on board.

This consensus, however, does not run very deep. And now that everyone is an environmentalist, we will have to talk about disagreements within the environmental movement. But over some fundamental issues the movement has always been divided. I want to address one of the most contentious and enduring disagreements in this epilogue.

Every environmental proposal is debated and fought, but this is not necessarily a sign of fundamental disagreement. There will always be arguments at the margins, where groups try to defend their private interests or where parties disagree about whether a particular measure costs too much or benefits too little. But some disagreements go deeper and are more general. One of these is expressed today in the Wise Use Movement, a reaction against more radical elements of the environmental movement that some people believe have too heavily influenced the regulatory laws and captured the agencies. The Wise Use Movement charges that these environmentalists are too absolutist and unreasonable; too much concerned with protecting wilderness, wetlands, and endangered species; and too little concerned with protecting the endangered economy and with promoting human welfare.

This particular debate is simply the most recent manifestation of a deeper split within the environmental movement, one with a long history. For several decades philosophers and nature writers have been debating preservationism and conservationism. Preservationists tend to want to protect certain regions and to enact land-use policies that would prohibit all commercial use and other disruptive human activities. They want rivers and wetlands to be not just safe but pristine. They want all species protected. They speak for Nature.

Conservationists would like to see the harmful environmental effects of commercial and other human activities regulated and, if possible, eliminated, but not the activities themselves. They appreciate Nature as a source of recreation for hikers, hunters, and fishermen, but they see as well the benefits of the minerals we mine and the timber we cut and the development of land. These are all different values that our environmental policies need to balance. The conservationist idea is nicely captured by the slogan prominently emblazoned on the wooden signs that greet visitors to our national forests today: Land of Many Uses.

Preservationists think the correct attitude toward Nature is to see it as having its own or intrinsic value. Conservationists see the value of Nature in its contribution to human welfare. Is Nature to be valued for itself or for its usefulness?

A century ago, as Steven Kelman points out in Part 1 of this volume, this same debate was exemplified in the lives and writings of John Muir and Gifford Pinchot, two colorful figures and giants of the early environmental movement in America. Muir was a Scottish immigrant who abandoned his family and left his industrial job in Wisconsin to explore the American wilderness by foot, traveling first through the Southeast but then heading West to the Sierra Nevada mountains in California. He experienced a personal epiphany in the beauty of America's wilderness, which he described in his diaries and later in articles he wrote for popular journals. Muir thought the American wilderness was truly unique. He described the breathtaking meadows of the Yosemite region, for example, as the "floweriest piece of world I've ever walked, one vast level, even flower bed, a sheet of flowers."

Already in the 1870s Muir feared that the economic development of the American West threatened the natural beauty of the area. He would write articles extolling the wilderness, trying to bring to public attention the need to protect it. He founded the Sierra Club for this purpose, and he devoted his life to helping others appreciate America's natural wonders and generating political support for government programs to protect them. "God over the ages had saved trees from natural disasters," he would write, "but could not save them from fools. Only Uncle Sam can do that."

Gifford Pinchot, whom Kelman describes as one of the great public managers in American history, was the first head of the United States Forest Service. Pinchot was also concerned about the impact of economic development on the vast areas of publicly owned land in America, but his concerns were different from Muir's. Pinchot feared that unregulated activity would lead to over-exploitation of the land's resources, and he advocated regulation to prevent this from happening. "There are just two things on this material earth," he wrote, "people and natural resources." Pinchot saw the need to manage the use of natural resources so that they would sustain their yield and support the economic growth of future generations of Americans. Pinchot was also sometimes brutally frank about the nature of his concern for the environment. "The first duty of the human race," he wrote, "is to control the earth it lives on." And he announced to the Society of American Foresters, "The object of our forest policy is not to preserve the forests because they are beautiful . . . or because they are refuges for the wild creatures of the wilderness . . . but . . . [for] the making of prosperous homes."

Today, conservationists are also concerned with the effects of environmental pollution on human health, and many of them are sympathetic to the recreational value of natural resources. The conservationist will support protectionist policies for these purposes, but these are simply natural extensions of Pinchot's original outlook.

The conflict between preservationists and conservationists raises some genuinely difficult issues about the value of Nature, and it raises some issues about how we understand the nature of our values. My aim is not to resolve this conflict but to try to shed some light on how we should understand it. In order to make any progress here, however, we will need a more precise description of the difference between preservationism and conservationism.

Preservationists, as I have claimed, maintain that the environment, or Nature, has intrinsic value and ought to be protected for its own sake. We ought to act in ways that show respect for Nature. Now, if conservationism is a position that represents a different or contrasting view, what could it mean? I think it could be interpreted to mean either of two different things, and perhaps both. One way to deny that Nature has intrinsic value is to insist that only human beings, or only creatures capable of reasoning or suffering, have intrinsic moral value. Humanity is the source of all moral worth, and Nature or anything else has moral value only through its connection to human goodness or ideals of human life. Without humanity, nothing has any moral worth whatsoever. According to this interpretation, conservationists are anthropocentric about moral values, and preservationists reject anthropocentrism.

A second way of denying that Nature has intrinsic value is to insist that it has only instrumental moral worth or that it is valuable only as a means. An effective anesthetic has value instrumentally because it is a means to something else and ultimately to something that has intrinsic value, in this case the relief of suffering. If sentient creatures became immune to pain, anesthetics would cease to have any value at all. On this interpretation, conservationists hold that Nature has only instrumental value. Our only concern should be to promote human good—human flourishing, excellence, or happiness—and the relief of suffering. The environment is simply an instrument for doing that. This is its value. Preservationists, again, insist that Nature is not only a means to human welfare but that it is itself a valuable thing.

It is common to regard conservationists as committed to both anthropocentrism about the source of moral values and instrumentalism regarding the value of Nature. It seems natural to link these positions, because it may seem that if humanity is the source of all moral value, then Nature can have only instrumental value insofar as it promotes human interests. But it is important to distinguish these two concepts and treat them separately. In the moral arguments that involve anthropocentrism and instrumental versus intrinsic value, the policy maker must be clear as to how these concepts are defined and related before policy decisions are made. As the essays in this volume have shown, it is not enough to assume standard definitions and connections—one must argue their case. This brings us to policy decision making.

The most critical way of characterizing the conflict between preservationism and conservationism focuses on their different implications for how we think about and decide environmental policies. For the conservationist, conflicts between environmental values and other values can be seen as problems of resource allocation. The goal is to find the best balance or distribution of resources, the one that maximizes human flourishing and minimizes suffering. The technical details of these decisions will be enormously complex and subject to errors, but in principle at least these decisions seem well suited to techniques and methods like risk analysis, decision analysis, and cost-benefit analysis. Some conservationists, at least, do not object to trying to translate the values of Nature into economic terms, trying to put a dollar value on protecting wilderness or wetlands, in order to weigh these values and compare them to other and competing claims on our resources. Some go further and argue that unless we are willing to see environmental values in economic terms they will not be weighed seriously in policy decisions. If we value Nature, we should show our willingness to pay for it.

Preservationists reject this way of thinking about Nature. We can respect the intrinsic value of Nature, they argue, only if we come to see conflicts between environmental values and other values not merely as competing claims on resources or inputs into an economic optimization problem but as conflicts between fundamental and incommensurable values. Just as we reject putting a price on human life or human rights in order to decide what resources to allocate to police protection, the judicial system, or emergency health care, so we must reject the idea that a simple algorithm or analytical technique can tell us how to balance the costs of environmental protection against other costs. We must learn to act in ways that show respect for Nature. We should refuse to think of it as an economic commodity or as an investment waiting to be cashed in or exchanged.

I have interpreted the conflict between preservationists and conservationists in terms of anthropocentrism, instrumentalism, and the suitability of methods like cost-benefit analysis. Earlier chapters in this book have considered the first two in detail, so I will concentrate on the third.

The undeniable truth of conservationism is that we have to make trade-offs. Each incremental reduction of air pollution costs more money. At what point should we stop? When does it become reasonable to tolerate some level of pollution and to spend the money on something else—on creating jobs, for example, or on improving schools, or even on improving health care? The conservationist urges us to face these questions explicitly. Cost-benefit analysis is one decision-making method for helping us to do this. It asks us to find a way of comparing our different needs and values and weighing their importance. It suggests that we use money as the metric, because the policy questions are usually about how much money we should spend or whether the benefits of some proposed action are worth its costs.

But preservationism also tells us something about the value of Nature—the kind of intrinsic value it has, or the way that Nature enriches or adds meaning to our lives. Preservationism forces us to recognize something about the nature of values, that is, as Kelman puts it, they can occupy different domains and trigger different kinds of appropriate responses in us. Economic commodities, as Marx pointed out, are fully appreciated when we understand their use value and their exchange value. But there are other kinds of value, too.

Some objects have sacred value; some have symbolic value. These objects have meaning for us apart from their usefulness. They have a semantic element, so to speak, which involves our cognitive capacities. This is why we talk about some values as requiring appreciation and some objects as demanding respect. Things can have exchange and use values for animals. A squirrel will take the acorn that it hopes to eat, recognizing its use value, but it will also drop it to escape a predator, thus revealing its exchange value.

The other domains of value are available only to rational creatures, those capable of reason and understanding. The preservationist, in opposing methods like cost-benefit analysis for making environmental decisions, suggests that forcing us to think about Nature as an economic commodity is incompatible with, and a threat to, its intrinsic value. The preservationist wants to protect the environment not only from pollution and development; she wants also to protect its meaning or its domain of value.

Think about love or friendship. The commitment of love is a commitment to the non-exchangeability of the person you love. It is a commitment to respond, in ways large and small, without calculating. Why buy a birthday present for my wife? Why not give her money instead? She enjoys shopping, and I don't. I will give her money for her gift and some more for her time. And I will explain how we both come out ahead. It is an efficient transaction. But it is impossible to express love in such a way. It is an inappropriate response, as she would certainly point out to me.

And yet the non-exchangeability of the object of love and the unique set of actions appropriate to expressing love do not necessarily mean that love is an incommensurable value. Part of our rationality allows us to stand back and reflect on our lives, including the role of love in it. We can make some kind of assessment of its importance. We can perhaps vaguely understand its worth to us and compare it to the worth of our other values. At the same time, if these reflective thoughts become too dominant in our minds, they can color our first-order, nonreflective attitudes and perhaps destroy them. There is no logical or conceptual impossibility in assessing and comparing different values, even sacred or symbolic values. We can take some measure of their importance in our lives. But there may be a psychological impossibility involved. It may be that forcing us to think in certain ways about some kinds of values will destroy the special meaning they have for us, destroy our ability to respond appropriately to them.

The preservationist, as I have said, is sensitive to the different domains of value and the need to protect the integrity of these domains. This is not a unique feature of the value of Nature. Consider, for example, the appeals of public radio stations for our contributions and support. Sometimes these stations pitch the economic value of public radio. How much is this programming worth to you? For thirty-five dollars, you get public radio for less than ten cents a day. Isn't that a bargain for no commercials and the programming you want?

I think this kind of appeal is basically misguided. It leads us to think about public radio in ways that make us less interested in supporting it. It does not take a genius to realize that if thirty-five dollars is a bargain for public radio, then three dollars and sixty-five cents will bring you the same benefits for only a penny a day. And the best bargain would be to get them for free (John Gillroy shows us the dynamics of these strategic situations in Part 1 of this book).

But public radio also makes a different appeal. These voices appeal to fairness, and they talk about listeners doing their part to support a unique and special kind of broadcasting. They appeal to our communitarian instincts. We are invited to join the station, to become a member. This is more effective fund-raising, I believe, because it asks us to think of the value of public radio differently. It asks us to support a community of which we are a part, not to buy a commodity off the shelf.

Of course, we have to make trade-offs. Even sacred values involve economic costs, which can become prohibitive. I will ask you to think about one final, unpleasant example. We cannot afford to fully protect the citizens of Baltimore from assault. We cannot afford to provide the effective security that would eliminate rapes, muggings, and domestic abuse in our midst. How much should we be spending on this kind of security? I cannot answer that question, but I will ask you to imagine someone taking a certain kind of rational approach to answering it. Given our other needs and the limited resources we have available to us, it is possible that we are spending too much on security. But how can we think like this? How can we think that, given our limited resources, we do not have the optimal number of rapes or muggings in our cities? No public figure could talk like that. Nor can we maintain our decency or humanity if we are inclined to think about these basic human values in this way. This is the kind of threat to which preservationists respond when they hear about optimal pollution levels or optimal mixes of land use. Their concern is not just for the decay of the environment or the threat to the planet; it is for the threat to our humanity as well. And yet, trade-offs have to be made.

As we struggle with these issues, we must also realize that there are justifications for supporting a preservationist position on environmental policies that have nothing to do with the nature of the moral values involved. I will simply mention two of them. First, there are psychological effects and biases associated with different ways of framing an issue. Some recent studies in the psychology of preference, for example, show that different ways of framing a choice question lead most people to focus on different aspects of the choices involved and to weigh these aspects differently. This can lead to different preferences or to preference reversals when the same choice is described differently. When subjects are asked simply to express a preference between two alternatives, for instance, between cleaner air and some consumer good, they will usually focus on the attributes of the alternatives that are supported by reasons and arguments, and they will form their preferences accordingly. When they are asked to rate the same alternatives on a common scale, however, as for instance by assigning an economic value or expressing a willingness to pay for them, most subjects will focus on other aspects and reverse their preferences. So the preservationist can be seen as fighting a certain kind of bias when an issue is framed in a particular way. When the preservationist says that we need to think about Nature differently, she could mean that if we framed the issues in terms of choice and not in terms of willingness to pay, we would express a different set of preferences for the alternatives.

Alternatively, one can be a preservationist—or a conservationist—for reasons of political strategy. The environmental movement has become politically quite sophisticated over the years, and its leaders realize that laws or policies written in an absolutist language do not mean that our practices will become absolutist too. Compromises will always be made, and these environmentalists may reason that the final result will be better and more balanced if their position remains absolutist. They need not fear that the forces of business and development or that citizen groups like the Wise Use Movement will disappear, leading us away from a reasonable middle ground.

I have been focusing only on the moral aspects of the conflict between preservationism and conservationism, not on the other, important non-moral aspects. One can claim, as most of the authors in this book have, that Nature has intrinsic value, and a special kind of intrinsic value, without losing sight of the basic policy trade-offs that must be made in our social lives. Preservationists tell us that we have to think about Nature in a special way; conservationists remind us that we must also balance the value of Nature against other values.

Because we are all environmentalists now, the conflict will continue within the environmental movement. I think this is appropriate, for the conflict is also within each of us individually. It points to a psychological disharmony or a tension between the demands of rationality and the demand to be true to the integrity of our values.

When I am hiking in the woods, I sometimes manage to step into a different domain. Unbothered by my normal concerns, I find myself contemplating Nature. I am bedazzled by the delicate and complex balance of living and nonliving things surrounding me. But then, sometimes, a different image strikes me, and I sense the precariousness of this natural harmony. I see a ceaseless competition for dominance being played out over decades and millennia. Competition, as much as harmony, is an image of the divine sentiment, and it is also an image of ourselves.

We are marvelous in similar ways, and our own psyches are also balanced precariously. Our competing reasons, our different feelings and moral intuitions, all coexist in a struggle for dominance. How this conflict works itself out, through our collective choices, will affect our long-term viability as both individuals and citizens of political communities.

About the Book and Editor

Public decisions on environmental risk have traditionally been weighed in terms of the principle of efficiency and its methodologies, such as cost-benefit and risk-benefit analysis. These original essays argue for moving beyond the market paradigm toward making policy that incorporates environmental values. Scholars representing a broad range of disciplines present a thorough analysis and methodological investigation of environmental risk and the potential for integrating environmental values into the policymaking process. They address the normative and theoretical roots of environmental risk, describe the distinct domain that exists for environmental values as opposed to economic values, and look at the conflicts between economic and environmental values within the applied context of the NIMBY (not in my back yard) phenomenon.

John Martin Gillroy is assistant professor of political science and director of the Public Policy Studies Program at Trinity College.

Index